The
Moon Almanac
2020

ISBN 13   978-0-9911266-7-5

email   info@proximopress.com

Website   www.themoonalmanac.com
              www.proximopress.com

PROXIMO PRESS

Data used in *The Moon Almanac* and *The Moon Calendar* was provided by the US Naval Observatory and other sources of astronomical information. The contents were produced with the guidance of Larry Sessions, astronomer.

CONTENTS

Welcome!

With a presidential election certain to command attention in 2020, the focus on politics may overshadow most other topics for most of the year. The Moon will be hard pressed to compete with the combative environment that accompanies such elections, must less such a one as this one is certain to be, unless it becomes part of a debate on astral influences. Despite generations of progressive thinking grounded in science, the Moon still holds a place of prestige among those who believe humans — as well as plants, pets, and machinery — are subject to forces from the stars, constellations, and the Moon.

Astrology, which was originally a partner in the development of modern astronomy, has retained its core beliefs despite a widening gap between the philosophy, methodology, and tools of science and the application of astrological principles. Not a complete divorce, however, as astrology has adopted computerization as an effective tool. And since nothing in science, hard or soft, has provided an explanation for the political circus that has consumed the United States since the last presidential election, astrology can do no worse.

Also coming in 2020 is a calendar year with a blue moon, another opportunity to mix science and humbug. As reported widely in astronomical publications over the past few decades, the concept of the blue moon is nothing more than a curiosity imposed by a juxtaposition of the natural cycles of the Earth, Moon, and Sun with the artificial constriction of the civil calendar. That is, by an accident of coincidence, every few years there is a calendar month that begins with one full moon and ends with another. This happens just as frequently with first quarter moon, last quarter moons, and new moons, but people pay little or no attention to the first two, and cannot see the third (lunar calendars traditionally do not start at a new moon, but the first sighting of a new crescent moon).

Hundreds of years ago, when western societies were largely illiterate and calendar conventions were controlled by Christian churches, a season, marked by three months — and three full moons — was occasionally upset by the appearance of a fourth full moon. This was usually not seen as a good sign and the errant full moon was designated as a "blue" moon, because in earlier forms of English ("blewe," for example), the adjective did not refer to the familiar color, but a sad or sorrowful feeling or condition. The irregular appearance of a full moon in an otherwise orderly procession was considered a bad sign.

In 2020, the month of October will have two full moons, one on the first and the other on the thirty-first. Its the second one that gets the label blue moon, and as it happens, it falls only three days in advance of the presidential election. With or without this calendrical curiosity, there is already enough ill will, venom, and dark thoughts to make this election a highly charged event. Whatever the outcome, we can expect that one group or another will look for appropriate targets to blame if the outcome is not in their favor, the blue moon being one.

Meanwhile, calendars, lunar cycles, and full moons will continue as always. We hope there is enough of interest in this volume to keep readers intrigued throughout the year, and to appreciate the methodical rhythms that mark the punctual passages of our nearest celestial neighbor.

TIME ZONES

See Time Conversions chart on page 6.

PST	MST	CST	EST	UT
Pacific Standard Time	Mountain Standard Time	Central Standard Time	Eastern Standard Time	Universal Standard Time
−8 hours	−7 hours	−6 hours	−5 hours	

DAYLIGHT SAVING TIME: SUBTRACT 1 ADDITIONAL HOUR

- DAYLIGHT SAVING TIME BEGINS March, 2020
- DAYLIGHT SAVING TIME ENDS November 1, 2020

LUNATIONS

A lunation is a single lunar month, from new moon to new moon. This basic unit of calendar time, observed and used by cultures around the world for thousands of years, was not standardized by astronomers until the twentieth century. An agreement by major countries established the current lunation series as beginning on January 16, 1923. In 2020, the series continues with **Lunation 1200**, which began with the new moon of December 26, 2019, and runs through 13 cycles to **Lunation 1212**, beginning with the new moon on December 14, 2020.

- Average lunation   29 days 12 hours 44 minutes
- Shortest lunation   29 days 6 hours 35 minutes
- Longest lunation   29 days 19 hours 55 minutes
- Average phase   7.38 calendar days
- Extremes phases   6–9 calendar days

THE MOON'S ILLUMINATION

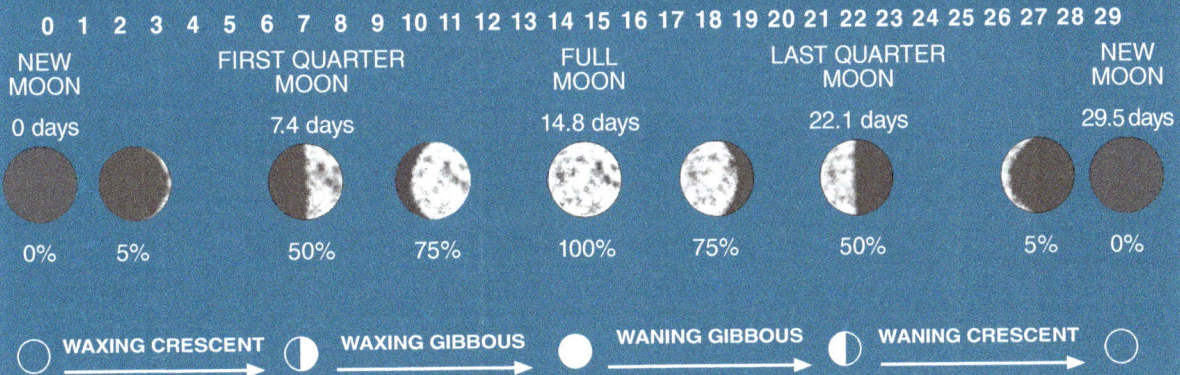

0 1 2 3 4 5 6 7 8 9 10 11 12 13 14 15 16 17 18 19 20 21 22 23 24 25 26 27 28 29

NEW MOON		FIRST QUARTER MOON		FULL MOON		LAST QUARTER MOON		NEW MOON	
0 days		7.4 days		14.8 days		22.1 days		29.5 days	
0%	5%	50%	75%	100%	75%	50%	5%	0%	

WAXING CRESCENT →   WAXING GIBBOUS →   WANING GIBBOUS →   WANING CRESCENT →

UT UNIVERSAL TIME (24 HRS)	UK CIVIL TIME	EST EASTERN STANDARD TIME −5 hours	CST CENTRAL STANDARD TIME −6 hours	MST MOUNTAIN STANDARD TIME −7 hours	PST PACIFIC STANDARD TIME −8 hours
0:00	12:00 AM	7:00 PM	6:00 PM	5:00 PM	4:00 PM
1:00	1:00 AM	8:00 PM	7:00 PM	6:00 PM	5:00 PM
2:00	2:00 AM	9:00 PM	8:00 PM	7:00 PM	6:00 PM
3:00	3:00 AM	10:00 PM	9:00 PM	8:00 PM	7:00 PM
4:00	4:00 AM	11:00 PM	10:00 PM	9:00 PM	8:00 PM
5:00	5:00 AM	12:00 AM	11:00 PM	10:00 PM	9:00 PM
6:00	6:00 AM	1:00 AM	12:00 AM	11:00 PM	10:00 PM
7:00	7:00 AM	2:00 AM	1:00 AM	12:00 AM	11:00 PM
8:00	8:00 AM	3:00 AM	2:00 AM	1:00 AM	12:00 AM
9:00	9:00 AM	4:00 AM	3:00 AM	2:00 AM	1:00 AM
10:00	10:00 AM	5:00 AM	4:00 AM	3:00 AM	2:00 AM
11:00	11:00 AM	6:00 AM	5:00 AM	4:000 AM	3:00 AM
12:00	12:00 PM	7:00 AM	6:00 AM	5:00 AM	4:00 AM
13:00	1:00 PM	8:00 AM	7:00 AM	6:00 AM	5:00 AM
14:00	2:00 PM	9:00 AM	8:00 AM	7:00 AM	6:00 AM
15:00	3:00 PM	10:00 AM	9:00 AM	8:00 AM	7:00 AM
16:00	4:00 PM	11:00 AM	10:00 AM	9:00 AM	8:00 AM
17:00	5:00 PM	12:00 PM	11:00 AM	10:00 AM	9:00 AM
18:00	6:00 PM	1:00 PM	12:00 PM	11:00 AM	10:00 AM
19:00	7:00 PM	2:00 PM	1:00 PM	12:00 PM	11:00 AM
20:00	8:00 PM	3:00 PM	2:00 PM	1:00 PM	12:00 PM
21:00	9:00 PM	4:00 PM	3:00 PM	2:00 PM	1:00 PM
22:00	10:00 PM	5:00 PM	4:00 PM	3:00 PM	2:00 PM
23:00	11:00 PM	6:00 PM	5:00 PM	4:00 PM	3:00 PM

TIME ZONE CONVERSIONS — PREVIOUS DAY

MOON DISTANCE

APOGEE

The Moon orbits around the Earth on an elliptical path, an oval that puts it closer to us at some points in the lunar month. The closest point of this ellipse is the perigee; the point when it is farthest is the apogee.

The distance between these two extremes is about 10 percent, enough to create a visible difference in the size of the full moon — as much as 14 degrees wider if it falls at one extreme or the other. This effect is compounded when the full moon is rising or setting; the closeness to the horizon and the proximity to trees, buildings, and other objects combine to produce this illusion.

On April 7 (in all U.S. time zones) the Moon will be at its closest to Earth for the year, at 1:10 PM EST. The closest together that a perigee and a full or new moon will come in 2020 is on October 16, when the perigee occurs at 6:45 AM EST and the new moon occurs about six hours later, at 12:31 PM EST. The closest approach of the Moon for the year is on April 7, with the full moon a day later, making this date the largest and brightest of the year.

PERIGEE				APOGEE			
				Jan. 2*	1:31 AM	251,393 mi	404,578 km
Jan. 13	8:22 PM	227,399 mi	365,963 km	Jan. 29	9:29 PM	251,897 mi	405,389 km
Feb. 10	8:32 PM	223,981 mi	360,463 km	Feb. 26	11:36 AM	252,448 mi	406,276 km
Mar. 10*	6:34 AM	221,905 mi	357,122 km	● Mar. 24	3:24 PM	252,704 mi	406,688 km
● Apr. 7	6:10 PM	221,772 mi	356,908 km	April 20	7:02 PM	252,563 mi	406,461 km
May 6*	3:05 AM	223,479 mi	359,655 km	May 18*	7:46 AM	252,018 mi	405,583 km
June 3*	3:38 AM	226,406 mi	364,365 km	June 15	12:58 AM	251,404 mi	404,596 km
June 30*	2:10 AM	229,259 mi	368,957 km	July 12	7:28 PM	251,158 mi	404,200 km
July 25*	4:55 AM	228,892 mi	368,366 km	Aug. 9	1:52 PM	251,442 mi	404,657 km
Aug. 21	11:00 AM	225,876 mi	363,512 km	Sep. 6*	6:32 AM	252,031 mi	405,605 km
Sep. 18	1:45 PM	223,122 mi	359,080 km	Oct. 3	5:24 PM	252,475 mi	406,319 km
Oct. 16	11:48 PM	221,775 mi	356,912 km	Oct. 30	6:47 PM	252,520 mi	406,392 km
Nov. 14	11:49 AM	222,350 mi	357,838 km	Nov. 27	12:30 AM	252,208 mi	405,890 km
Dec. 12	8:43 PM	224,797 mi	361,776 km	Dec. 24	4:33 PM	251,661 mi	405,009 km

● Closest of the year. ● Farthest of the year. *Occurs the previous day in one or more North American time zones.

2020 MOON PHASES

	NEW MOON	FIRST QUARTER	FULL MOON	LAST QUARTER
JANUARY Lunation 1200		3 4:45 AM UT 2 11:45 PM EST 10:45 PM CST 9:45 PM MST 8:45 PM PST	10 7:21 PM UT 2:21 PM EST 1:21 PM CST 12:21 PM MST 11:21 AM PST WOLF MOON	17 12:58 AM UT 16 7:58 PM EST 6:58 PM CST 5:58 PM MST 4:48 PM PST
FEBRUARY Lunation 1201	24 9:42 PM UT 4:42 PM EST 3:42 PM CST 2:42 PM MST 1:42 PM PST	2 1:42 AM UT 1 8:42 PM EST 7:42 PM CST 6:42 PM MST 5:42 PM PST	9 7:33 AM UT 2:33 AM EST 1:33 AM CST 12:33 AM MST 8 11:33 PM PST SNOW MOON	15 10:17 PM UT 5:17 PM EST 4:17 PM CST 3:17 PM MST 2:17 PM PST
MARCH Lunation 1202	23 3:32 PM UT 10:32 AM EST 9:32 AM CST 8:32 AM MST 7:32 AM PST	2 7:57 PM UT 2:57 PM EST 1:57 PM CST 12:57 PM MST 11:57 AM PST	9 5:48 PM UT 12:48 PM EST 11:48 AM CST 10:48 AM MST 9:48 AM PST WORM MOON	16 7:34 AM UT 2:34 AM EST 1:34 AM CST 12:34 AM MST 15 11:34 PM PST
	24 9:28 AM UT 4:28 AM EST 3:28 AM CST 2:28 AM MST 1:28 AM PST			

Daylight Saving Time begins March 8. Add one hour where in effect.

SPRING EQUINOX
March 20 3:49 AM UT
19 10:49 PM EST
9:49 PM CST
8:49 PM MST
7:49 PM PST

2020 MOON PHASES

NEW MOON	FIRST QUARTER	FULL MOON	LAST QUARTER	
	1 10:21 AM UT	8 2:35 AM UT	14 10:56 PM UT	**APRIL**
	5:21 AM EST	7 9:35 PM EST	5:56 PM EST	Lunation 1203
	4:21 AM CST	8:35 PM CST	4:56 PM CST	
	3:21 AM MST	7:35 PM MST	3:56 PM MST	
	2:21 AM PST	6:35 PM PST	2:56 PM PST	
		PLANTER'S MOON		
23 2:26 AM UT	30 8:38 PM UT	7 10:45 AM UT	14 2:03 PM UT	**MAY**
22 9:26 PM EST	3:38 PM EST	5:45 AM EST	9:03 AM EST	Lunation 1204
8:26 PM CST	2:38 PM CST	4:45 AM CST	8:03 AM CST	
7:26 PM MST	1:38 PM MST	3:45 AM MST	7:03 AM MST	
6:26 PM PST	12:38 PM PST	2:45 AM PST	6:03 AM PST	
		MILK MOON		
22 5:39 PM UT	30 3:30 AM UT	5 7:12 PM UT	13 6:24 AM UT	**JUNE**
12:39 PM EST	29 10:30 PM EST	2:12 PM EST	1:24 AM EST	Lunation 1205
11:39 AM CST	9:30 PM CST	1:12 PM CST	12:24 AM CST	
10:39 AM MST	8:30 PM MST	12:12 PM MST	12 11:24 PM MST	
9:39 AM PST	7:30 PM PST	11:12 AM PST	10:24 PM PST	
		STRAWBERRY MOON		
21 6:41 AM UT	28 8:16 AM UT			
1:41 AM EST	3:16 AM EST			
12:41 AM CST	2:16 AM CST			
20 11:41 PM MST	1:16 AM MST			
10:41 PM PST	12:16 AM PST			

SUMMER SOLSTICE
June 20 9:43 PM UT
4:43 PM EST
3:43 PM CST
2:43 PM MST
1:43 PM PST

9

2020 MOON PHASES

NEW MOON	FIRST QUARTER	FULL MOON	LAST QUARTER

JULY
Lunation 1206

		5 4:44 AM UT	**12** 11:29 PM UT
		4 11:44 PM EST	6:29 AM EST
		10:44 PM CST	5:29 AM CST
		9:44 PM MST	4:29 AM MST
		8:44 PM PST	3:29 AM PST
		BUCK MOON	

AUGUST
Lunation 1207

20 5:33 PM UT	**27** 12:33 PM UT	**3** 3:59 PM UT	**11** 4:45 PM UT
12:33 PM EST	7:33 AM EST	10:59 AM EST	11:45 AM EST
11:33 AM CST	6:33 AM CST	9:59 AM CST	10:45 AM CST
10:33 AM MST	5:33 AM MST	8:59 AM MST	9:45 AM MST
9:33 AM PST	4:33 AM PST	7:59 AM PST	8:45 AM PST
		CORN MOON	

SEPTEMBER
Lunation 1208

19 2:42 AM UT	**25** 5:58 PM UT	**2** 5:22 AM UT	**10** 9:26 AM UT
18 9:42 PM EST	12:58 PM EST	12:22 AM EST	4:26 AM EST
8:42 PM CST	11:58 AM CST	**1** 11:22 PM CST	3:26 AM CST
7:42 PM MST	10:58 AM MST	10:22 PM MST	2:26 AM MST
6:42 PM PST	9:58 AM PST	9:22 PM PST	1:26 AM PST
		BARLEY MOON	

17 11:00 AM UT	**24** 1:55 AM UT		
6:00 AM EST	**23** 8:55 PM UT		
5:00 AM CST	7:55 PM UT		
4:00 AM MST	6:55 PM UT		
3:00 AM PST	5:55 PM UT		

FALL EQUINOX
September 23   1:30 PM UT
8:30 AM EST
7:30 AM CST
6:30 AM MST
5:30 AM PST

2020 MOON PHASES

NEW MOON	FIRST QUARTER	FULL MOON	LAST QUARTER

Daylight Saving Time ends November 1. Roll clocks back one hour where in effect.

		1 9:05 PM UT	**10** 12:40 AM UT	**OCTOBER**
		4:05 PM EST	**9** 7:40 PM EST	Lunation 1209
		3:05 PM CST	6:40 PM CST	
		2:05 PM MST	5:40 PM MST	
		1:05 PM PST	4:40 PM PST	
		HARVEST MOON		

16 7:31 PM UT	**23** 1:23 PM UT	**31** 2:49 PM UT	**8** 1:46 PM UT	**NOVEMBER**
2:31 PM EST	12:23 PM EST	9:49 AM EST	8:46 AM EST	Lunation 1210
1:31 PM CST	11:23 AM CST	8:49 AM CST	7:46 AM CST	
12:31 PM MST	10:23 AM MST	7:49 AM MST	6:46 AM MST	
11:31 AM PST	9:23 AM PST	6:49 AM PST	5:46 AM PST	
		HUNTER'S MOON		

15 5:07 AM UT	**22** 4:45 AM UT	**30** 9:30 AM UT	**8** 12:37 AM UT	**DECEMBER**
12:07 AM EST	**21** 11:45 PM EST	4:30 PM EST	**7** 7:37 PM EST	Lunation 1211
11:07 PM CST	10:45 PM CST	3:30 PM CST	6:37 PM CST	
10:07 PM MST	9:45 PM MST	2:30 PM MST	5:37 PM MST	
9:07 PM PST	8:45 PM PST	1:30 PM PST	4:37 PM PST	
		BEAVER MOON		

14 4:16 PM UT	**21** 11:41 PM EST	**30** 3:28 AM UT	
11:16 AM EST	6:41 PM EST	**29** 10:28 PM EST	
10:16 AM CST	5:41 PM EST	9:28 PM CST	
9:16 AM MST	4:41 PM EST	8:28 PM MST	
8:16 AM PST	3:41 PM EST	7:28 PM PST	
		COLD MOON	

WINTER SOLSTICE
December 21 10:02 AM UT
5:02 AM EST
4:02 AM CST
3:02 AM MST
2:02 AM PST

11

MOON HORIZONS

The Moon follows roughly the same path as the Sun across the sky — a virtual line known as the ecliptic. Because of the ecliptic's tilt relative to the Earth, the Moon appears to rise — and set — at different points along the horizon from night to night and from season to season. This point varies according to location, especially between cities far apart from north to south.

The chart opposite shows rising and setting points of the full moon for representative U.S. cities, measured by the azimuth (horizontal point on the horizon). North is 0 degrees, East is 90 degrees, South is 180 degrees, and West is 270 degrees (as illustrated at right). Imaginary lines rising from these points reach up to the zenith, the point directly overhead.

From one full or new moon to the next, the point at which it intersects the horizon can jump by as much as 10 degrees or more. But at two points during the year — the summer and winter solstices, in June and December, respectively — this "jump" decreases dramatically, an effect created by the unique position of the Earth on the ecliptic at these key seasonal points. The same effect holds true for the Sun.

Once above the horizon, the Moon transits overhead across the local meridian, a virtual line running from north to south. The meridian marks the halfway point between the east and west horizons and, because the Moon is always to the south at the point when it transits, traditional almanacs called this the Moon's "southing" or "moon up."

The Moon's transit is different for every location — it crosses over New York City at a different time than it crosses over Cincinnati. The transit at any given spot is about 49 minutes later from one day to the next. The only two constants that are easy to figure out about the transit are during the new moon and the full moon. At new moon, the Moon is directly in line with the Sun — the transit at this phase is the same as the Sun's, about midday. At full moon, the Moon is opposite the sun — the transit at this phase is about halfway through the night.

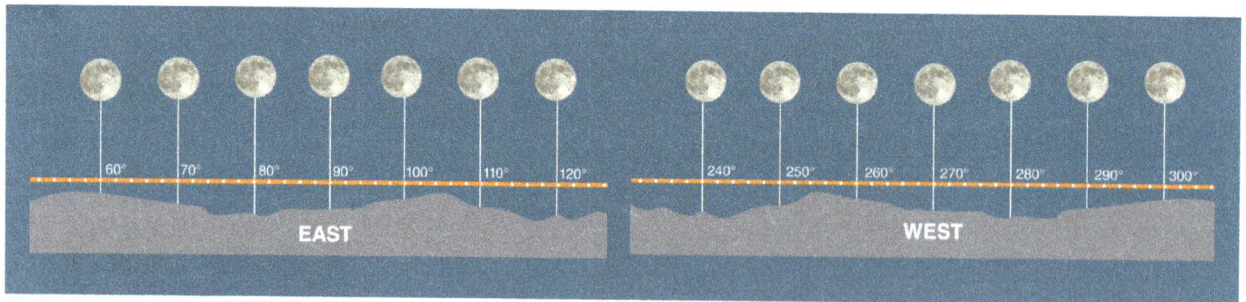

EAST 60° 70° 80° 90° 100° 110° 120°

WEST 240° 250° 260° 270° 280° 290° 300°

FULL MOON RISE/SET HORIZON LOCATOR*

		JAN 10	FEB* 9	MAR 9	APR 7	MAY 7	JUN 5	JUL 4	AUG 3	SEP* 2	OCT 1	OCT 31	NOV 30	DEC 29
Atlanta	RISE	62°	72°	82°	93°	110°	117°	119°	114°	100°	90°	75°	63°	60°
	SET	298°	291°	282°	271°	253°	245°	241°	244°	257°	267°	283°	296°	300°
Boston	RISE	58	69	80	93	113	121	123	117	102	91	73	60	56
	SET	302	294	283	272	251	242	236	240	255	266	284	299	304
Chicago	RISE	59	69	81	93	113	121	123	117	101	90	73	60	56
	SET	302	293	283	271	251	242	237	241	255	266	284	299	304
Columbus	RISE	60	70	81	93	112	119	121	116	101	90	74	61	57
	SET	301	293	283	271	252	243	238	242	255	266	284	298	303
Dallas	RISE	63	72	82	93	110	117	119	113	100	90	75	63	60
	SET	298	290	281	271	253	245	241	244	257	267	283	295	299
Denver	RISE	60	70	81	93	112	120	122	116	107	90	73	60	57
	SET	301	292	282	271	252	243	238	242	250	267	284	298	303
Miami	RISE	65	73	82	244	108	115	116	112	100	90	76	65	62
	SET	296	289	281	117	255	247	243	246	258	267	282	293	297
Los Angeles	RISE	62	67	82	93	11	117	120	113	105	90	74	63	60
	SET	298	295	281	270	253	245	241	244	252	267	283	296	300
Nashville	RISE	61	71	81	93	11	118	120	114	101	90	74	62	59
	SET	299	291	282	271	253	244	240	243	256	266	283	296	301
New York City	RISE	59	70	81	93	112	120	122	116	101	91	74	60	57
	SET	301	293	283	271	252	243	237	241	255	266	284	298	303
Phoenix	RISE	62	72	82	93	110	117	119	113	105	90	74	63	60
	SET	298	290	281	271	253	245	241	244	252	267	283	296	300
Salt Lake City	RISE	59	70	81	94	113	120	123	116	107	90	73	60	59
	SET	301	292	282	271	251	242	238	242	250	267	284	299	302
San Francisco	RISE	61	66	82	94	112	119	121	115	106	90	73	61	58
	SET	300	296	282	270	252	243	239	243	251	267	284	297	302
Seattle	RISE	55	61	81	94	116	125	127	119	109	90	71	56	52
	SET	305	301	284	270	249	238	233	238	247	266	287	303	308
Washington DC	RISE	60	70	81	93	111	119	121	116	101	91	74	61	58
	SET	300	292	283	271	252	244	238	242	250	266	283	297	302

* Dates of full moons are in EST. *Dates vary in some time zones because of Daylight Saving Time (not noted here).

MOON HEIGHT

The Moon closely follows the ecliptic, the same path that the Sun appears to take across the sky. This path is tilted relative to the Earth's axis, making the height of the Sun and Moon vary throughout the year. During a full moon, when the Moon is opposite the Sun, it will appear high in the sky when the Sun is low, and vice versa.

In a 24-hour period, as the Earth rotates into a different position relative to the ecliptic, an observer will see a marked difference in height between the Sun and the Moon when they are in opposition. This recurring pattern varies a bit because the Moon's path "wanders" back and forth across the ecliptic.

Sometimes it is above it, sometimes below. When the two coincide during a full or new moon, the result is an eclipse.

The apparent altitude of the Moon varies according to an observer's location. The chart below indicates the variations in peak altitude of the full moon in 2020. Halfway between moon rise and moon set on the day of a full moon, this point occurs about midnight, a point referred to as its transit. The time and the actual altitude vary according to longitude and latitude (the figures used on this page are computed for Denver, Colorado).

Full Moon at its Highest Point in the Night Sky in 2020

Date	Altitude
JAN 10	73°
FEB 9	68°
MAR 9	61°
APR 7	52°
MAY 7	37°
JUN 5	30°
JUL 4	26°
AUG 3	28°
SEP 1	33°
OCT 1	46°
OCT 31	59°
NOV 30	70°
DEC 28	75°

EARLIEST CRESCENT MOON

At several times throughout the year, during the new moon, the Moon is not only close to the Sun, but it overlaps the Sun's position, producing a solar eclipse. When there is no eclipse event, the Moon is off to one side or the other of the Sun, but the light from the Sun makes it impossible to see from the Earth.

But every day after the new moon, the position of the Moon moves about 13 degrees farther away from the Sun. Within a few days after the new moon, the Sun has set and the sky begins to darken while the first crescent of the Moon is still above the horizon, producing a highly noticeable sight in the western sky.

For millenia, the point in time when this first crescent moon can be seen with the naked eye marked the beginning of a new lunar month. The Islamic religion, for one, still depends on the sighting of the first crescent moon to officially begin a monthly calendar cycle.

With the aid of computers and hundreds of years of experience, the prediction of when the first crescent moon can be seen has become more exact, but complex factors keep this from being a certainty. Factors affecting the visibility of a young moon include the presence of obstructions on the local horizon (e.g. hills, mountains, trees), clouds or haze, the position of the Moon relative to the Sun (high or low relative to the ecliptic), and the time of year (the angle between the ecliptic and the horizon varies from season to season). Latitude is also a factor; middle latitudes are less favored while high and low latitudes yield better results. Higher altitudes up the odds as well.

At best, observers without the use of binoculars, telescopes, or solar filters have been able to see a first

FIRST CRESCENT VIEWING OPPORTUNITIES IN 2020

These dates and approximate times are for locations in the Eastern Standard Time zone.

March 24	15 hours
April 22	23 hours
June 21	19 hours
August 19	24 hours

crescent moon when it is only about fifteen hours "old" — that is, fifteen hours after the exact time of the new moon. Most of the time, casual observers will not see an early crescent moon until two or more days after the new moon.

The potential to see any first crescent moon when it is less than twenty-four hours old is low. There are only a few times during the year when the timing is right to produce an optimal viewing "window." The prime conditions for early first crescent observation occur when the Moon is closest to Earth (at the perigee of its orbit) and coincides within a few days of the date of a new moon. The spring months, when the ecliptic is at its steepest angle relative to the horizon at sunset, provide the most favorable time of year.

A first crescent moon at thirty-two hours old, photographed from the window of a passenger plane flying between Phoenix and Denver.

SKY SIGHTS NEAR THE MOON IN 2020

The following dates represent prime viewing events for planets and major stars visible to the naked eye and near the Moon in 2019. For many of these occasions, the viewing opportunities may extend additional days before and after the date listed, particularly with the major planets — the specific angles and dates mark the closest distance between the object and the Moon.

When a bright star or planet is 1° or less from the Moon (or another star or planet), the event is called an occultation. Because of differences in location for viewers, the time of this closest approach may be during daylight hours — and therefore not visible — or the event may not be observable at all.

In the convention of astronomical directions, *up* or *above* are officially north, *down* is south, *right* is east, and *left* is to the west when an observer is facing the Moon. The illustrations represent approximate positions projected in advance by Stellarium, an online planetarium program. The time of day or night is referenced for North America unless specially noted.
www.stellarium.com

JANUARY

20 Mars 2° south of waning crescent moon.
23 Jupiter less than 1° north of new moon.
28 Venus 4° north of waxing crescent moon.

FEBRUARY

2 Vesta less than 1° south of first quarter moon.
18 Mars less than 1° south of waning crescent moon.
19 Jupiter less than 1° north of waning crescent moon.
20 Saturn about 2° north of waxing crescent moon.
27 Venus 6° north of waxing crescent moon.

MARCH

1 Vesta less than 1° north of first crescent moon.
18 Mars less than 1° north and Jupiter between 1°–2° north of waning crescent moon.
19 Saturn 2° north of waning crescent moon.
28 Venus 7° north of waxing crescent moon.
29 Vesta less than 1° north of waxing crescent moon.

APRIL

14 Pluto a little more than 1° north and Jupiter 2° north of last quarter moon.
15 Saturn 2° north of waning crescent moon.
16 Mars 2° north of waning crescent moon.

26 Vesta less than 1° south of waxing crescent moon.
26 Venus 6° north of waxing crescent moon (two days before Venus' greatest illumination of year).

MAY

12 Jupiter 2° north and Saturn 3° north of waxing crescent moon.
15 Mars 3° north of waning crescent moon.
24 Venus 4° north and Vesta less than 1° south of first crescent moon.

JUNE

8 Jupiter 2° north of waning gibbous moon.
9 Saturn 3° north of waning

Use your fingers and hand to approximate measurements at a distance. The little finger at the and of an outstretched arm equals about 1 degree (about twice the angular diameter of the Moon or Sun). The first and second fingers spread apart represent about 5 degrees and a fist, the width of the hand, is about 10 degrees.

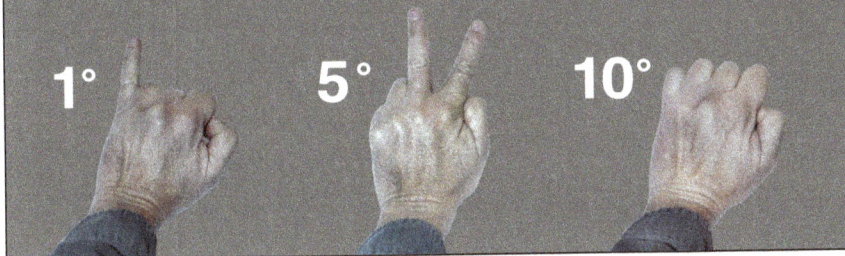

1°   5°   10°

gibbous moon.
13 Mars 3° north of last crescent moon.
19 Venus less than 1° south of waning crescent moon.

JULY

5 Penumbral lunar eclipse; Jupiter about 2° north of full moon.
6 Saturn 2° north of waning gibbous moon.
11 Mars 2° north of nearly last crescent moon.
17 Venus 3° south of waning crescent moon.

AUGUST

2 Jupiter between 1°–2° north and Saturn 2° north of

almost full moon.
9 Mars less than 1° north of waning gibbous moon.
15 Venus 4° south of waning crescent moon.
29 Jupiter between 1°–2° north and Saturn 2° north of waxing gibbous moon.

SEPTEMBER

6 Mars less than 1° south of waning gibbous moon.
14 Venus 4° south of waning crescent moon.
25 Jupiter between 1°–2° north and Saturn 2° north of waxing gibbous moon.

OCTOBER

3 Mars less than 1° north of

waning gibbous moon.
14 Venus 4° south of waning crescent moon.
22 Jupiter 2° north of waxing crescent moon.
23 Saturn 3° north of first quarter moon.
29 Mars 3° north of almost full moon.

NOVEMBER

12 Venus 3° south of waning crescent moon.
19 Jupiter 2° north and Saturn 3° north of waxing crescent moon.
25 Mars 5° north of waxing gibbous moon.

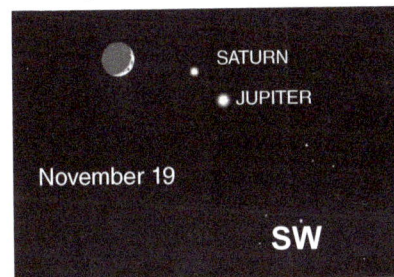

SATURN
JUPITER
November 19
SW

DECEMBER

12 Venus less than 1° south of waning crescent moon.
17 Jupiter 3° north of Moon; Saturn 3° north of early crescent moon.
23 Mars 6° north of waxing gibbous moon.

2020 MOON RISE AND MOON SET

City		JANUARY 2	JANUARY 10	JANUARY 16	FEBRUARY 1	FEBRUARY 9	FEBRUARY 15	MARCH 2	MARCH 9	MARCH 16
Atlanta	RISE	12:29 PM	5:49 PM	12:35 AM	11:54 AM	6:53 PM	12:38 AM	11:37 AM	6:50 PM	1:42 AM
	SET	—	7:35 AM	12:29 PM	12:22 AM	7:55 AM	11:42 AM	1:03 AM	7:08 AM	11:52 AM
Boston	RISE	11:37 AM	4:27 PM	—	10:50 AM	5:40 PM	—	10:20 AM	5:47 PM	1:12 AM
	SET	11:47 PM	7:05 AM	11:27 AM	—	7:19 AM	10:31 AM	12:28 AM	6:24 AM	10:31 AM
Chicago	RISE	11:44 AM	4:38 PM	—	10:58 AM	5:51 PM	12:05 AM	10:30 AM	5:57 PM	1:19 AM
	SET	11:56 PM	7:13 AM	11:35 AM	—	7:26 AM	10:40 AM	12:36 AM	6:31 AM	10:41 AM
Columbus	RISE	11:25 PM	4:25 PM	11:32 AM	11:42 AM	5:36 PM	11:42 PM	10:16 AM	6:39 PM	1:54 AM
	SET	—	6:47 AM	11:18 AM	11:22 PM	7:02 AM	10:25 AM	12:11 AM	7:10 AM	11:29 AM
Dallas	RISE	12:19 PM	5:43 PM	12:27 PM	11:46 AM	6:46 PM	12:29 PM	11:30 AM	6:43 PM	1:31 AM
	SET	—	7:42 AM	12:20 PM	12:13 AM	7:45 AM	11:35 AM	12:53 AM	6:58 AM	11:46 AM
Denver	RISE	11:54 AM	4:58 PM	12:04 AM	11:12 AM	6:09 PM	12:14 AM	10:47 AM	6:12 PM	1:25 AM
	SET	—	7:18 AM	11:48 AM	—	7:32 AM	10:56 AM	12:43 AM	6:39 AM	11:00 AM
Miami	RISE	12:10 PM	5:51 PM	12:14 AM	11:45 AM	6:47 PM	12:10 AM	11:35 AM	6:38 PM	1:05 AM
	SET	—	6:58 AM	12:17 PM	—	7:24 AM	11:37 AM	12:30 AM	6:43 AM	11:54 PM
Nashville	RISE	11:39 AM	4:52 PM	—	11:01 AM	5:59 PM	—	10:41 AM	5:58 PM	12:58 AM
	SET	11:53 PM	6:51 AM	11:37 AM	—	7:10 AM	10:48 AM	12:18 AM	6:21 AM	10:56 AM
Los Angeles	RISE	11:47 AM	5:10 PM	—	11:12 AM	5:03 PM	12:00 AM	10:55 AM	6:12 PM	1:04 AM
	SET	—	6:56 AM	11:47 AM	—	6:28 AM	11:01 AM	12:25 AM	6:28 AM	11:12 AM
New York City	RISE	11:49 AM	4:45 PM	—	11:04 AM	5:56 PM	12:06 AM	10:37 AM	6:01 PM	1:18 AM
	SET	—	7:12 AM	11:41 AM	—	7:27 AM	10:47 AM	12:36 AM	6:34 AM	10:49 AM
Phoenix	RISE	12:22 PM	5:45 PM	12:31 AM	11:48 AM	6:49 PM	12:33 AM	11:31 AM	6:47 PM	1:36 AM
	SET	—	7:29 AM	12:22 PM	12:17 AM	7:49 AM	11:36 AM	12:58 AM	7:02 AM	11:48 AM
Salt Lake City	RISE	12:23 PM	5:23 PM	12:34 AM	11:38 AM	6:36 PM	12:45 AM	11:12 AM	6:40 PM	1:57 AM
	SET	—	7:50 AM	12:16 PM	12:23 AM	8:03 AM	11:22 AM	1:14 AM	7:09 AM	11:25 AM
San Francisco	RISE	12:05 PM	5:16 PM	12:16 AM	11:25 AM	5:12 PM	12:24 AM	11:03 AM	6:27 PM	1:31 AM
	SET	—	7:24 AM	12:01 PM	12:04 AM	6:55 AM	11:11 AM	12:50 AM	6:49 AM	11:19 AM
Seattle	RISE	12:07 PM	4:41 PM	12:22 AM	11:11 AM	4:43 PM	12:43 AM	10:34 AM	6:18 PM	2:06 AM
	SET	—	7:59 AM	11:51 AM	12:15 AM	7:24 AM	10:49 AM	1:18 AM	7:01 AM	10:43 AM
Wash., DC	RISE	12:00 PM	5:03 PM	12:07 AM	11:19 AM	6:13 PM	12:15 AM	10:54 AM	6:15 PM	1:25 AM
	SET	—	7:19 AM	11:55 AM	—	7:35 AM	11:03 AM	12:44 AM	6:44 AM	11:07 AM

2020 MOON RISE AND MOON SET

		APRIL 1	APRIL 7	APRIL 14	APRIL 30	MAY 7	MAY 14	MAY 29	JUNE 5	JUNE 13	JUNE 28
Atlanta	RISE	11:57 AM	6:47 PM	1:27 AM	11:50 AM	7:56 PM	1:36 AM	11:53 AM	7:53 PM	1:12 AM	1:05 PM
	SET	1:49 AM	6:15 AM	11:33 AM	1:29 AM	6:00 AM	12:17 PM	12:56 AM	5:17 AM	12:55 PM	12:45 AM
Boston	RISE	10:34 AM	5:53 PM	12:59 AM	10:31 AM	7:18 PM	1:01 AM	10:42 AM	7:23 PM	12:25 AM	12:10 PM
	SET	1:21 AM	5:21 AM	10:10 AM	12:59 AM	4:51 AM	11:03 PM	12:19 AM	4:00 AM	11:54 AM	11:53 PM
Chicago	RISE	10:45 AM	6:03 PM	1:06 AM	10:41 AM	7:27 PM	2:08 AM	11:52 AM	7:31 PM	12:32 AM	12:19 PM
	SET	1:28 AM	5:29 AM	11:21 AM	1:05 AM	5:00 AM	12:13 PM	1:26 AM	4:10 AM	12:03 PM	12:01 AM
Columbus	RISE	11:32 AM	6:43 PM	1:41 AM	11:28 AM	8:03 PM	1:44 AM	11:37 AM	8:05 PM	1:11 AM	1:00 PM
	SET	2:02 AM	6:10 AM	11:08 AM	1:40 AM	5:44 AM	11:58 AM	1:02 AM	4:56 AM	12:45 PM	12:41 AM
Dallas	RISE	11:51 AM	6:39 PM	1:16 AM	11:44 AM	7:47 PM	1:25 AM	11:47 AM	7:43 PM	1:01 AM	12:57 PM
	SET	1:38 AM	6:06 AM	11:27 AM	1:18 AM	5:52 AM	12:10 PM	12:45 AM	5:11 AM	12:47 PM	12:36 AM
Denver	RISE	11:04 AM	6:15 PM	1:11 AM	11:01 AM	7:35 PM	1:14 AM	11:09 AM	7:37 PM	12:41 AM	12:32 PM
	SET	1:33 AM	5:40 AM	10:40 AM	1:10 AM	5:15 AM	11:30 AM	12:32 AM	4:27 AM	12:17 PM	12:11 AM
Miami	RISE	11:59 AM	6:27 PM	12:50 AM	11:50 AM	7:25 PM	1:04 AM	11:48 AM	7:17 PM	12:48 AM	12:47 PM
	SET	1:12 AM	5:57 AM	11:35 AM	12:53 AM	5:53 AM	12:13 PM	12:25 AM	5:17 AM	12:42 PM	12:26 AM
Nashville	RISE	11:00 AM	5:57 PM	12:44 AM	10:54 AM	7:11 PM	12:51 AM	10:59 AM	7:10 PM	12:23 AM	12:15 PM
	SET	1:06 AM	5:25 AM	10:36 AM	12:45 AM	5:06 AM	11:22 AM	12:10 AM	4:22 AM	12:03 PM	11:55 PM
Los Angeles	RISE	11:17 AM	6:09 PM	12:49 AM	11:11 AM	7:19 PM	12:56 AM	11:15 AM	7:18 PM	12:01 AM (12)	12:27 PM
	SET	1:11 AM	5:34 AM	10:53 AM	12:50 AM	5:19 AM	11:37 AM	12:16 AM	4:37 AM	11:20 AM	12:04 AM
New York C.	RISE	10:52 AM	6:05 PM	1:06 PM	10:48 AM	7:27 PM	1:09 AM	10:57 AM	7:30 PM	12:35 AM	12:22 PM
	SET	1:27 AM	5:33 AM	10:28 AM	1:05 AM	5:06 AM	11:19 AM	12:27 AM	4:17 AM	12:07 PM	12:05 AM
Phoenix	RISE	11:53 AM	6:43 PM	1:21 AM	11:46 AM	7:52 PM	1;29 AM	11:50 AM	7:48 PM	12:35 AM (12)	1:01 AM
	SET	1:43 AM	6:08 AM	11:29 AM	1:23 AM	5:54 AM	12:12 PM	12:49 AM	5:12 AM	11:55 AM	12:38 AM
Salt Lake C.	RISE	11:30 AM	6:45 PM	1:43 PM	11:26 AM	8:06 PM	1:44 AM	11:36 AM	8:09 PM	1:10 AM	1:01 PM
	SET	2:05 AM	6:08 AM	11:06 AM	1:42 AM	5:41 AM	11:57 AM	1:03 AM	4:53 AM	12:45 PM	12:39 AM
San Fran.	RISE	11:23 AM	6:28 PM	1:17 PM	11:19 AM	7:44 PM	1:21 AM	11:26 AM	7:44 PM	12:23 AM (12)	12:45 PM
	SET	1:39 AM	5:51 AM	10:59 AM	1:17 AM	5:30 AM	11:47 AM	12:40 AM	4:45 AM	11:33 AM	12:21 AM
Seattle	RISE	10:46 AM	6:32 PM	1:53 PM	10:47 AM	8:09 PM	1:46 AM	11:05 AM	8:18 PM	12:39 AM (12)	12:46 PM
	SET	2:15 AM	5:51 AM	10:23 AM	1:49 AM	5:10 AM	11:23 AM	1:02 PM	4:14 AM	11:19 AM	12:24 AM
Wash., DC	RISE	11:11 AM	6:17 PM	1:12 AM	11:06 AM	7:36 PM	1:17 AM	11:14 AM	7:37 PM	12:46 AM	12:35 PM
	SET	1:34 AM	5:45 AM	10:47 AM	1:13 AM	5:21 AM	11:35 AM	12:36 AM	4:34 AM	12:21 PM	12:16 AM

2020 MOON RISE AND MOON SET

City		JULY 4	JULY 12	JULY 27	AUGUST 3	AUGUST 11	AUGUST 25	SEPTEMBER 2	SEPTEMBER 10	SEPTEMBER 23
Atlanta	RISE	8:38 PM	12:07 AM	1:11 PM	8:01 PM	11:31 AM	1:19 PM	7:41 PM	11:14 AM	1:22 PM
	SET	5:46 AM	12:33 PM	—	5:35 AM	1:09 PM	—	6:24 AM	1:51 PM	11:25 PM
Boston	RISE	8:09 PM	11:13 AM	12:27 PM	7:28 PM	10:52 PM	12:44 PM	6:56 PM	10:37 AM	12:53 PM
	SET	4:23 AM	11:40 PM	11:22 PM	4:16 AM	12:29 PM	10:33 PM	5:17 AM	1:22 PM	10:02 PM
Chicago	RISE	7:23 PM	12:20 AM	12:35 PM	7:34 PM	—	12:52 PM	7:03 PM	10:47 AM	1:01 PM
	SET	3:37 AM	12:49 PM	11:00 PM	4:26 AM	12:37 PM	10:43 PM	5:27 AM	1:29 PM	10:13 PM
Columbus	RISE	8:50 PM	12:02 AM	1:13 PM	8:10 PM	11:17 PM	1:27 PM	8:42 PM	—	1:35 PM
	SET	5:21 AM	12:29 PM	11:43 PM	5:13 AM	1:14 PM	—	6:10 AM	2:03 PM	—
Dallas	RISE	7:32 PM	11:58 AM	1:02 PM	7:50 PM	11:23 PM	1:09 PM	7:31 PM	11:08 PM	1:11 PM
	SET	4:43 AM	12:25 PM	11:46 PM	5:29 AM	12:59 PM	—	6:17 AM	1:40 PM	—
Denver	RISE	7:27 PM	12:01 AM	12:45 PM	7:40 PM	—	12:59 PM	6:44 PM	—	1:06 PM
	SET	3:56 AM	11:31 PM	11:13 PM	4:45 AM	12:45 PM	11:01 PM	4:41 AM	1:34 PM	10:32 PM
Miami	RISE	8:01 PM	11:50 PM	12:46 PM	7:28 PM	11:24 PM	12:46 PM	7:18 PM	—	12:44 PM
	SET	5:49 AM	12:14 PM	11:43 PM	5:36 AM	12:40 PM	—	6:16 AM	1:14 PM	11:28 PM
Nashville	RISE	7:00 PM	11:17 PM	12:24 PM	7:16 PM	11:09 PM	12:34 PM	6:53 PM	11:02 PM	12:39 PM
	SET	3:52 AM	11:44 AM	11:02 PM	4:39 AM	12:23 PM	10:55 PM	5:31 AM	1:08 PM	10:28 PM
Los Angeles	RISE	7:05 PM	11:25 PM	12:34 PM	7:21 PM	11:21 PM	12:41 PM	6:30 PM	11:18 PM	12:44 PM
	SET	4:09 AM	11:54 AM	11:13 PM	4:56 AM	12:30 PM	11:10 PM	4:47 AM	1:13 PM	10:45 PM
New York City	RISE	8:16 PM	11:25 PM	12:37 PM	7:35 PM	11:07 PM	12:52 PM	7:06 PM	10:55 PM	1:00 PM
	SET	4:41 AM	11:52 AM	11:06 PM	4:33 AM	12:38 PM	10:50 PM	5:32 AM	1:28 PM	10:20 PM
Phoenix	RISE	7:38 PM	12:00 AM	12:00 PM	7:54 PM	11:57 PM	1:14 PM	7:04 PM	11:54 PM	1:16 PM
	SET	4:45 AM	12:28 PM	11:48 PM	5:31 AM	1:03 AM	11:46 PM	5:22 AM	1:45 PM	11:21 PM
Salt Lake City	RISE	8:00 PM	11:59 PM	1:16 PM	8:11 PM	11:13 PM	1:31 PM	7:14 PM	11:31 PM	1:38 PM
	SET	4:22 AM	12:30 PM	11:41 PM	5:11 AM	1:16 AM	—	5:08 AM	3:06 PM	10:57 PM
San Francisco	RISE	7:33 PM	11:42 PM	12:56 PM	7:47 PM	11:01 PM	1:07 PM	6:52 PM	11:24 PM	1:12 PM
	SET	4:15 AM	12:13 PM	11:26 PM	5:04 AM	12:54 PM	11:18 PM	4:58 AM	1:41 PM	10:51 PM
Seattle	RISE	8:11 PM	11:41 PM	1:11 PM	8:14 PM	11:07 PM	1:36 PM	7:09 PM	10:48 PM	1:50 PM
	SET	3:38 AM	12:16 PM	11:15 PM	4:32 AM	1:15 PM	10:47 PM	5:36 AM	2:16 PM	10:12 PM
Wash., DC	RISE	8:22 PM	11:37 PM	12:47 AM	7:43 PM	11:23 PM	1:00 PM	7:16 PM	11:13 PM	1:07 PM
	SET	5:00 AM	12:04 PM	11:52 PM	4:51 AM	12:47 PM	11:08 PM	5:47 AM	1:35 PM	10:39 PM

20

2020 MOON RISE AND MOON SET

City		OCT 1	OCT 9	OCT 23	OCT 31	NOV 8	NOV 21	NOV 30	DEC 7	DEC 21	DEC 29
Atlanta	RISE	6:39 PM	11:35 PM	1:59 PM	6:05 PM	—	1:17 PM	5:50 PM	—	12:50 PM	5:18 PM
	SET	6:10 AM	1:34 AM	—	6:48 AM	1:50 PM	—	7:32 AM	1:04 PM	—	7:18 AM
Boston	RISE	5:46 PM	10:12 PM	1:30 PM	4:58 PM	11:16 PM	12:42 PM	4:30 PM	11:27 PM	12:01 PM	3:54 PM
	SET	5:11 AM	1:07 AM	10:53 PM	6:04 AM	1:16 PM	10:51 PM	7:00 AM	12:22 PM	11:45 PM	6:50 AM
Chicago	RISE	5:53 PM	10:22 PM	1:37 PM	5:06 PM	11:27 PM	12:48 PM	4:40 PM	11:37 PM	12:08 PM	4:04 PM
	SET	5:20 AM	1:14 AM	—	6:12 AM	1:23 PM	11:01 PM	7:07 AM	12:29 PM	11:55 PM	6:58 AM
Columbus	RISE	6:34 PM	—	2:11 PM	5:51 PM	—	1:25 AM	5:27 PM	—	12:48 PM	4:52 PM
	SET	6:02 AM	1:48 PM	—	6:50 AM	1:59 AM	11:46 PM	7:42 AM	1:07 PM	—	7:31 AM
Dallas	RISE	6:30 PM	—	1:48 PM	5:57 PM	—	11:57 PM	5:43 PM	—	12:40 PM	5:12 PM
	SET	6:01 AM	1:22 PM	— PM	6:39 AM	1:39 PM	1:06 PM	7:21 AM	12:54 PM	— AM	7:07 AM
Denver	RISE	6:03 PM	10:42 PM	1:41 PM	5:21 PM	11:45 PM	11:18 PM	4:58 PM	11:52 PM	12:17 PM	4:24 PM
	SET	5:33 AM	1:18 AM	11:23 PM	6:21 AM	1:28 PM	12:54 AM	7:13 AM	12:36 PM	—	7:02 AM
Miami	RISE	6:22 PM	11:38 PM	1:23 PM	5:58 PM	—	12:46 PM	5:51 PM	—	12:28 PM	5:21 PM
	SET	5:54 AM	12:55 AM	— AM	6:23 AM	1:17 PM	11:59 PM	6:57 AM	12:38 PM	— AM	6:40 AM
Nashville	RISE	5:49 PM	10:38 PM	1:16 PM	5:12 PM	10:36 PM	12:32 PM	4:53 PM	11:40 PM	12:01 PM	4:20 PM
	SET	5:19 AM	12:51 PM	11:17 PM	6:01 AM	1:05 PM	11:09 PM	6:48 AM	12:17 PM	11:53 PM	6:35 AM
Los Angeles	RISE	5:57 PM	10:55 PM	1:20 PM	5:23 PM	11:52 PM	12:37 PM	5:08 PM	11:53 PM	12:08 PM	4:37 PM
	SET	5:30 AM	12:55 PM	11:33 PM	6:09 AM	1:10 PM	11:24 PM	6:53 AM	12:23 PM	—	6:39 AM
New York C.	RISE	5:57 PM	10:30 PM	1:37 PM	5:13 PM	11:33 PM	12:50 PM	4:48 PM	11:41 PM	12:12 PM	4:12 PM
	SET	5:24 AM	1:13 PM	11:10 PM	6:14 AM	1:24 PM	11:07 PM	7:07 AM	12:31 PM	11:58 PM	6:57 AM
Phoenix	RISE	6:32 PM	11:31 PM	1:53 PM	5:59 PM	—	1:10 PM	5:45 PM	—	12:42 PM	5:13 PM
	SET	6:05 AM	1:28 PM	—	6:43 AM	1:43 PM	11:59 PM	7:26 AM	12:57 PM	—	7:12 AM
Salt Lake C.	RISE	6:31 PM	11:07 PM	2:13 PM	5:47 PM	—	1:25 PM	5:23 PM	—	12:46 PM	6:47 PM
	SET	6:02 AM	1:50 PM	11:49 PM	6:51 AM	1:59 PM	11:45 PM	7:45 AM	1:06 PM	—	9:19 AM
San Fran.	RISE	6:14 PM	11:01 PM	1:47 PM	5:35 PM	—	1:01 PM	4:15 PM	—	12:27 PM	4:43 PM
	SET	5:46 AM	1:24 PM	11:41 PM	6:31 AM	1:35 PM	11:35 PM	6:20 AM	12:45 PM	—	7:08 AM
Seattle	RISE	6:14 PM	10:23 PM	2:21 PM	5:17 PM	11:37 PM	1:25 PM	4:41 PM	11:55 PM	12:34 PM	4:04 PM
	SET	5:43 AM	2:02 AM	11:08 PM	6:46 AM	2:02 AM	11:12 PM	7:52 AM	12:59 PM	—	7:46 AM
Wash., DC	RISE	6:10 PM	10:49 PM	1:43 PM	5:28 PM	11:50 PM	12:58 PM	5:06 PM	11:56 PM	12:23 PM	4:31 PM
	SET	5:37 AM	1:19 PM	11:29 PM	6:24 AM	1:32 PM	11:23 PM	7:14 AM	12:41 PM	—	7:03 AM

Visible in northeastern and northwestern North America, Western Pacific Ocean, most of Australasia, Europe, Africa, eastern South America.

A BEGINS	Jan. 10	5:06 PM UT
		12:06 PM EST
		11:06 AM CST
		10:06 AM MST
		9:06 AM PST
B MIDDLE		7:10 PM UT
		2:10 PM EST
		1:10 PM CST
		12:10 PM MST
		11:10 AM PST
C ENDS		9:14 PM UT
		4:14 PM EST
		3:14 PM CST
		2:14 PM MST
		1:14 PM PST

Add 1 hour for Daylight Saving Time.

The graphics on pages 20–25 are derived from originals produced by NASA and the U.S. Naval Observatory. For more detailed information, visit the Eclipse Portal provided by the U.S. Naval Observatory: http://aa.usno.navy.mil/data/docs/UpcomingEclipses.php Eclipse Predictions by Fred Espenak, NASA GSFC Emeritus.

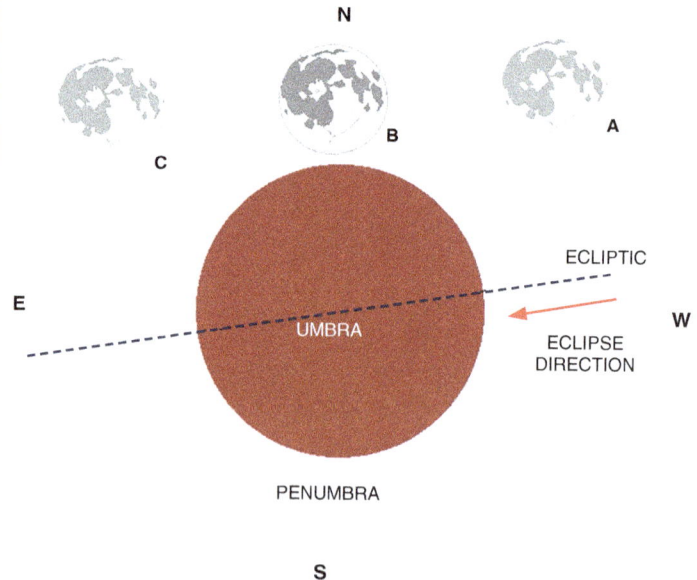

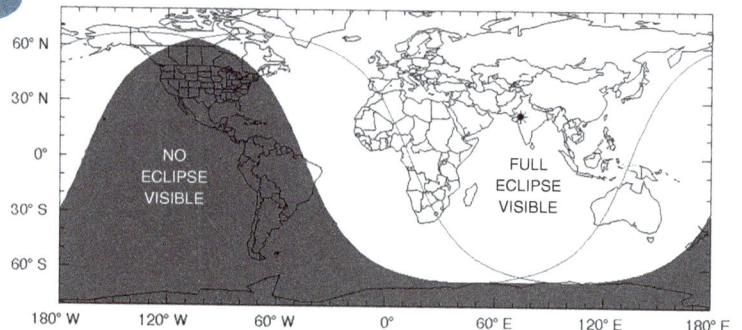

PENUMBRAL LUNAR ECLIPSE June 5, 2020

Not visible in North America. Visible in the western Pacific Ocean, Australasia, Asia (not including northern and eastern Russia), Antarctica, Europe (not including extreme north), Africa, eastern and southern South America.

A	BEGINS	June 5	5:43 PM UT
B	MIDDLE		7:25 PM UT
C	ENDS		9:07 PM UT

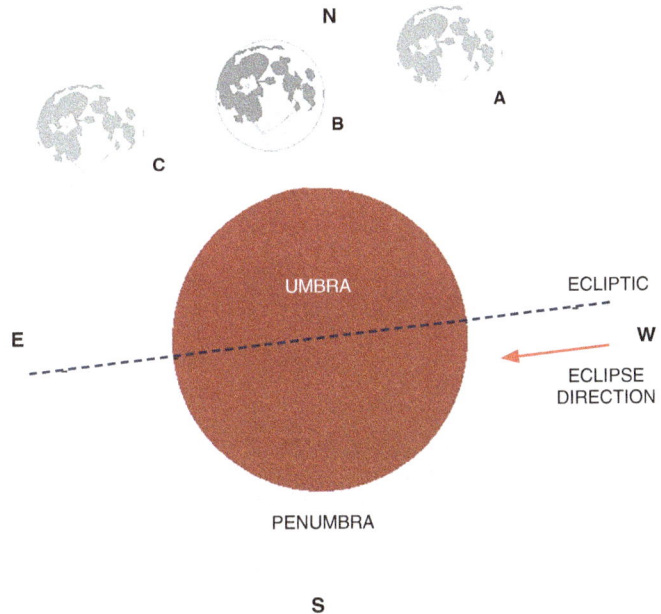

N

A

B

C

UMBRA

ECLIPTIC

E

W

ECLIPSE DIRECTION

PENUMBRA

S

The graphics on pages 19–21 are derived from originals produced by NASA and the U.S. Naval Observatory. For more detailed information, visit the Eclipse Portal provided by the U.S. Naval Observatory: http://aa.usno.navy.mil/data/docs/UpcomingEclipses.php
Eclipse Predictions by Fred Espenak, NASA GSFC Emeritus.

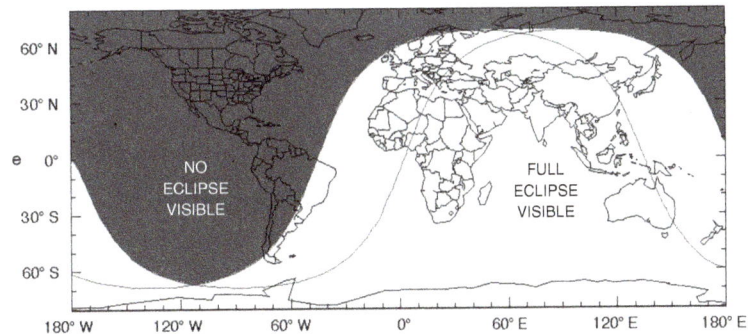

NO ECLIPSE VISIBLE

FULL ECLIPSE VISIBLE

ANNULAR ECLIPSE of the SUN June 21, 2020

Not visible in North America. Visible in Africa (not including western and southern areas), southeastern Europe, Middle East, Asia (not including northern and eastern Russia), Indonesia, Micronesia.

A	BEGINS	June 21	3:46 AM UT
B	MIDDLE		6:41 AM UT
C	ENDS		9:34 AM UT

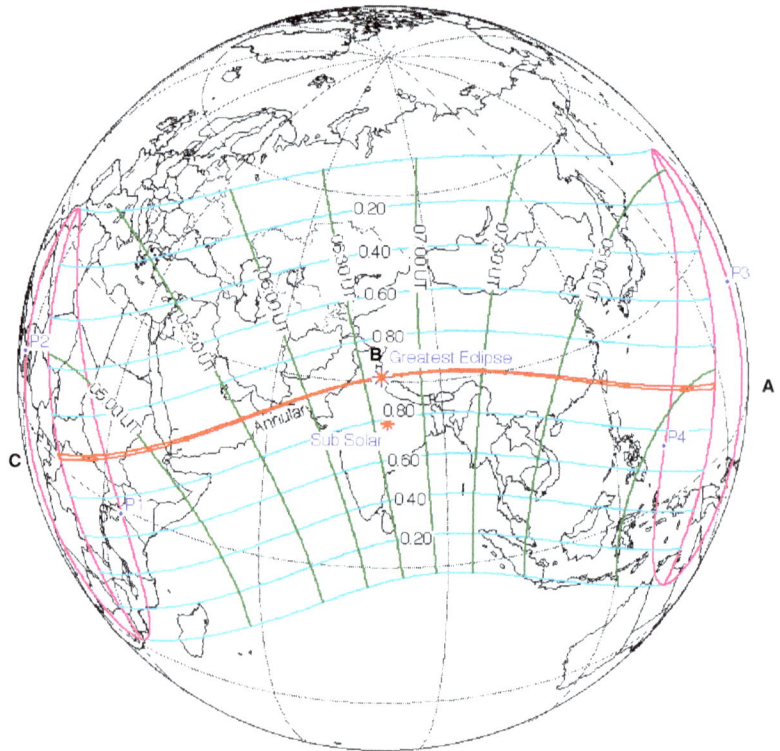

The graphics on pages 19–21 are derived from originals produced by NASA and the U.S. Naval Observatory. For more detailed information, visit the Eclipse Portal provided by the U.S. Naval Observatory: http://aa.usno.navy.mil/data/docs/UpcomingEclipses.php

Eclipse Predictions by Fred Espenak, NASA GSFC Emeritus.

PENUMBRAL LUNAR ECLIPSE July 5, 2020

Visible in Africa (not including northeastern areas), most of southern and western Europe, Antarctica, North America (not including northern areas), Central America, South America, most of Polynesia, New Zealand.

A BEGINS	July 5	3:09 AM UT
	July 4	10:09 PM UT
		9:09 PM EST
		8:09 PM CST
		7:09 PM MST
		6:09 PM PST
B MIDDLE	July 5	4:31 AM UT
	July 4	11:31 PM EST
		10:31 PM CST
		9:31 PM MST
		8:31 PM PST
C ENDS	July 5	5:54 AM UT
		12:54 AM EST
	July 4	11:54 PM CST
		10:54 PM MST
		9:54 PM PST

Add 1 hour for Daylight Saving Time.

The graphics on pages 19–21 are derived from originals produced by NASA and the U.S. Naval Observatory. For more detailed information, visit the Eclipse Portal provided by the U.S. Naval Observatory: http://aa.usno.navy.mil/data/docs/UpcomingEclipses.php
Eclipse Predictions by Fred Espenak, NASA GSFC Emeritus.

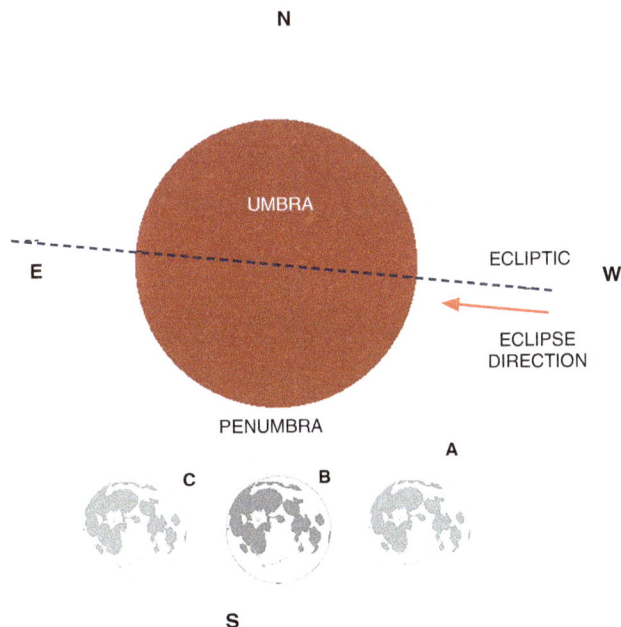

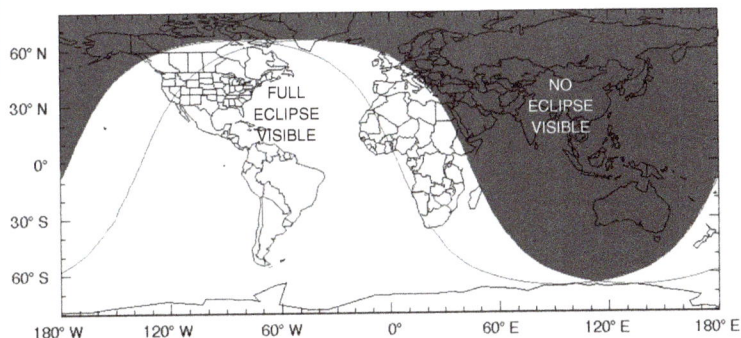

PENUMBRAL LUNAR ECLIPSE Nov. 30, 2020

Visible in northwestern Europe, North America, Central America, South America, Oceania, Australasia, most of Asia.

A BEGINS	Nov. 30	7:34 AM UT
		2:34 AM EST
		1:34 AM CST
		12:34 AM MST
	Nov. 29	11:34 PM PST
B MIDDLE	Nov. 30	9:34 AM UT
		4:34 AM EST
		3:34 AM CST
		2:34 AM MST
		1:34 AM PST
C ENDS		11:55 AM UT
		6:55 AM EST
		5:55 AM CST
		4:55 AM MST
		3:55 AM PST

Add 1 hour for Daylight Saving Time.

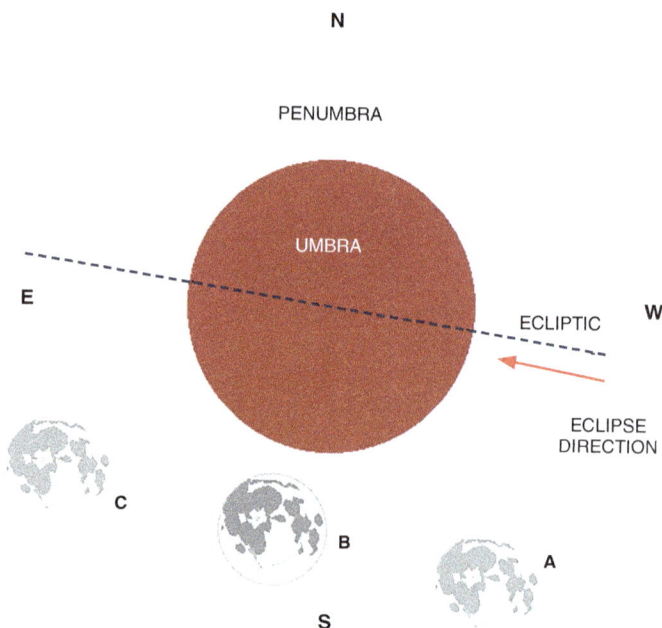

The graphics on pages 19–21 are derived from originals produced by NASA and the U.S. Naval Observatory. For more detailed information, visit the Eclipse Portal provided by the U.S. Naval Observatory: http://aa.usno.navy.mil/data/docs/UpcomingEclipses.php
Eclipse Predictions by Fred Espenak, NASA GSFC Emeritus.

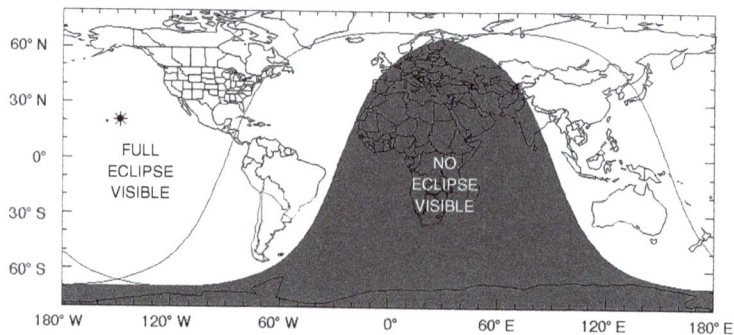

TOTAL ECLIPSE of the SUN December 14, 2020

Not visible in North America. Visible in the southern Pacific, South America (not including northern areas, parts of Antarctica, parts of southwestern Africa.

A BEGINS	Dec. 14	1:34 PM UT
B MIDDLE		4:13 PM UT
C ENDS		6:53 PM UT

Add 1 hour for Daylight Saving Time.

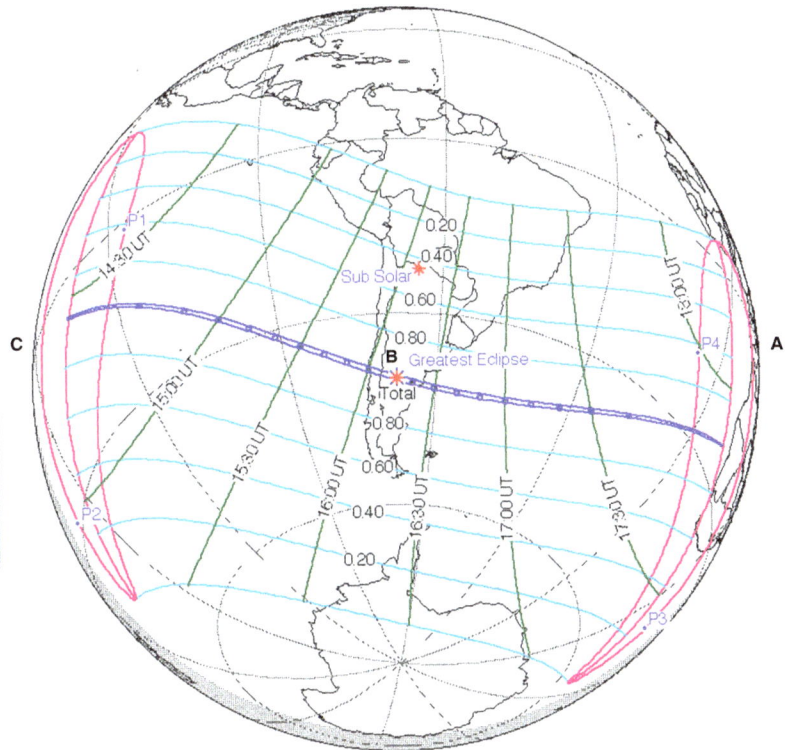

The graphics on pages 19–21 are derived from originals produced by NASA and the U.S. Naval Observatory. For more detailed information, visit the Eclipse Portal provided by the U.S. Naval Observatory: http://aa.usno.navy.mil/data/docs/UpcomingEclipses.php
Eclipse Predictions by Fred Espenak,
NASA GSFC Emeritus.

NAMES OF THE FULL MOONS

January 10 WINTER MOON or WOLF MOON. First full moon after the winter solstice (Dec. 21). The Dakota Sioux called it the Moon of the Terrible; to the Cherokee, it was the Month of the Cold Moon; to the Klamath it was Moon of the Little Finger's Partner.

February 9 SNOW MOON. Second full moon of the year, associated with the middle of winter. The Micmacs called it the Snow-Blinding Moon; Choctaw's called it the Little Famine Moon; the Osage referred to it as Light of Day Returns Moon.

March 9 WORM MOON. Last full moon before the spring equinox, when the first signs of warmer weather appear, or when the first worm castings are seen in the thawing earth. Colonists also called this the Fish Moon or the Sap Moon. Among native American names, it was Earth Cracks Moon to the Kutenai, the Deer Moon to the Natchez, and the Wind Strong Moon to the Taos Pueblo.

April 7 PLANTER'S MOON or EGG MOON. First moon after the spring equinox (March 19, US time zones), time for farmers to begin planting crops. The Algonquin called it the Pink Moon, referring to wild ground phlox (*phlox divaricata*), a native plant with a pink flower that bloomed during this time.

May 7 MILK MOON or FLOWER MOON. Fifth full moon of the year, relating to the season's first regular supply of milk as calves are weaned. The Lakota Sioux called this the Moon of the Shedding Ponies.

June 5 STRAWBERRY MOON. Sometimes interchanged with the name of the previous full moon, this was traditionally the season when strawberries came into season. The Cherokee called it the Month of the Green Corn Moon; the Choctaw called it the Windy Moon; the Osage called it the Buffalo Pawing Earth Moon; the Tlingit called it the Moon of the Salmon.

July 4 BUCK MOON. The first full moon after the summer solstice (June 20), also called the Summer Moon by colonists.

The Laguna called it the Corn Tassel Moon; the Ojibway called it the Raspberry Moon, and the Nez Perce called it the Red Salmon Time.

August 3 CORN MOON. Some colonists also called this the Woodcutter's Moon (time to begin gathering firewood for winter) or the Sturgeon Moon; the Choctaw called it the Women's Moon and the Haida called it the Collect Food for Winter Moon. In most years, it is the last full moon of summer.

September 2 BARLEY MOON. Usually called the Harvest Moon, but with an early date and the following full moon closest to the fall equinox, this event requires an alternate term (a traditional option, in any case), allowing the following full moons to maintain their traditional labels.

October 1 HARVEST MOON. The full moon closest to the fall equinox (September 22). At this time of year the moon rises at its steepest angle, seeming to linger near the horizon for several nights in a row, aiding farmers at harvest time. Also called the Moon When Wild Rice Is Stored for Winter Use (Dakotah Sioux).

October 31 HUNTER'S MOON. The full moon following the Harvest Moon. Also called the Blood Moon (referring to the harvesting of livestock or game), the Turkey's Moon (Natchez), and Falling of the Leaves Moon (Ojibway).

November 30 BEAVER MOON. Second full moon after the fall equinox (in some years, but not always), marking the season when beavers were fat and their pelts thick. Also called the Freezing Moon (Cheyenne), and the Sassafras Moon (Choctaw).

December 29 CHRISTMAS MOON or COLD MOON. Full moon closest to the winter solstice (December 21). Also called the Long Night Moon because the full moon rides high in the sky in this season. The Cheyenne called it the Big Freezing Moon and the Natchez called it the Bears Moon.

MOON HOLIDAYS in 2020

The lunar calendar, based on the 29.5-day lunar month, has been in use far longer than our current civil calendar, and parallels the development of various solar calendars. Lunar and lunisolar (a hybrid version combining elements of both lunar and solar years) calendar cultures include those of Mesopotamia, Babylon, ancient Egypt, ancient Greece, the Zoroastrian religion, the Chinese empire, ancient Celts, and others.

Some countries and a few religions still base their calendars on a lunar cycle, even though civil calendars may also be used. Islam and Judaism are among these. All of the Islamic countries use the official Muslim calendar that was first adopted in 634 AD; except for Orthodox sects, most Jews restrict their lunar observances to the major traditional holidays, where applicable.

Buddhists establish their new year with the lunar cycle, but variations abound. The first full moon in May is linked to the date of Buddha's Enlightenment and is the starting point of some of these calendars.

Countries that adhere to Mahayana Buddhism (including much of China, Mongolia, Japan, and Indonesia) celebrate their new year on the first full moon in January. Chinese New Year, the most familiar to Westerners, typically occurs between January 21 and February 21. Hindus in some regions recognize Ugadi, the first day of the month of Chaitra, as the beging of the new year. The first month of the lunar Hindu calendar, Chaitra begins with the new moon in March. In other regions, the occasion is recognized later, during the month of Vaisakha, which begins with the new moon in April. Other variations

2020 HOLIDAYS	
CHINESE NEW YEAR	January 25
NOROOZ (Persian New Year)	March 21
UGADI (Hindu New Year)	March 25
SAKA (Indian New Year)	March 21
THERAVADA (Buddhist New Year)	April 7
FIRST DAY OF PASSOVER	April 12
EASTER	April 21
SONGKRAN (Thai New Year)	April 13
WESAK (Buddha's Birthday)	May 7
FIRST DAY OF RAMADAN	April 24*
ROSH HASHANAH (Jewish New Year)	Sep. 19*
MUHARRAM (Islamic New Year)	Auigust 20*

* Begins at sunset the day before.

are common — scholars note at least thirty versions currently in use. In some, months are new moon to new moon; in others, full moon to full moon. Here is how a few of these holidays relate to the lunar cycle.

● **EASTER** The date for Easter in the Christian religion is designated as the first Sunday that falls after the first full moon (traditionally called the Pascal Full Moon) occurring on or after the spring equinox. The earliest date for Easter is March 22; the latest date is April 25. Easter shares its origins with Passover, and, because of the similarity in setting their dates, they often fall within about a week of each other. Occasionally,

Passover can be as much as one month later. The standard for Gregorian Easter (observed by most Western Christians) differs from Julian Easter (observed by orthodox sects). The dates can vary by one to four weeks because the calculations used by orthodox churches rely on the old Julian calendar, which was superceded by the Gregorian calendar in most of the Western world (in October 1582 AD).

The Chinese zodiac categorizes its traditional year according to one of twelve animals, each rotating through in a recurring cycle. In 2020, beginning on January 25, it will be the year of the Rat, the 12th symbol in the series.

- **MUHARRAM** Islam determines the start of its lunar-based calendar with the date the prophet emigrated from Mecca to Medina, (an event known as the Hijra), which was July 16, 622 AD on the Western calendar. In 2020, this day will be August 20. The first month of the Islamic calendar is Muharram; Al-Hijra Muharram is New Year's day. With all the Islamic lunar-based holidays, the dates move back from year to year until a thirty-two year (more or less) cycle matches them up once again with the Western calendar.

- **RAMADAN** Muslims mark the month in which the *Quran* was revealed to the prophet as their holiest holiday, Ramadan. This religious holiday begins with the sighting of the first crescent moon in the ninth month of the Muslim calendar, also known as Ramadan. The observances end with the sighting of the following crescent moon. The dates of the new moon marking Ramadan move backward relative to the civil calendar, by more than one week a year, synchronizing with the civil (solar) calendar about every thirty-two years.

- **ROSH HASHANAH** The first day of the month Tishrei in the Jewish calendar is celebrated as the Jewish New Year, Rosh Hashanah. It was originally marked by the first observance of the crescent moon, but visual evidence is no longer required. The date is adjusted according to when the new moon is calculated to occur — moving later by one or two days according to complex rules — and marked by a fixed number of days after the previous beginning of the year. Jewish year 5781 begins with Rosh Hashanah in 2020.

- **PASSOVER** (Pesach) is the traditional Jewish holiday designated to commemorate the exodus of the Jews from Egypt. This event occurred at the beginning of spring, and the celebration of Passover is configured to correspond to a lunar month near the spring equinox. In the fourth century, a system was designed to reduce the variables affecting this date. Passover begins on the fifteenth day of the Jewish month of Nisan — halfway through the lunar month and at the time of a full moon. Nisan is the first month after the spring (vernal) equinox, which can occur in either March or April. The earliest date for the beginning of Passover is March 21; the latest is April 20.

MOON LIGHT IN 2020

Each box below depicts the nighttime portion only of each twenty-four hour day. In each box, night begins at the bottom and proceeds upward, with daybreak at the top. The shaded bars indicate the relative amount of moonlight and when it is most concentrated within its nighttime period. During new moons, a box full of dark blue indicates no moonlight for the entire period; a box filled with white indicates a night when the full moon provides moonlight throughout. The total length of each night is given in hours and minutes; times can vary by a few minutes by location across similar latitudes but up to an hour or more across extreme latitudes (Boston compared to Miami, for example). This chart is based on New York City. Because hours of night and day are directly related to the movements of the Sun, they do not change by more than a minute or two from year to year for the same day of the year. The lunar cycle, however, shifts backwards (relative to the civil calendar) by 10 or more days a year, changing the periods when there is more moonlight.

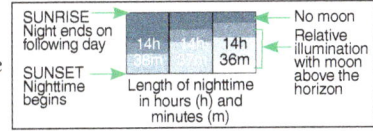

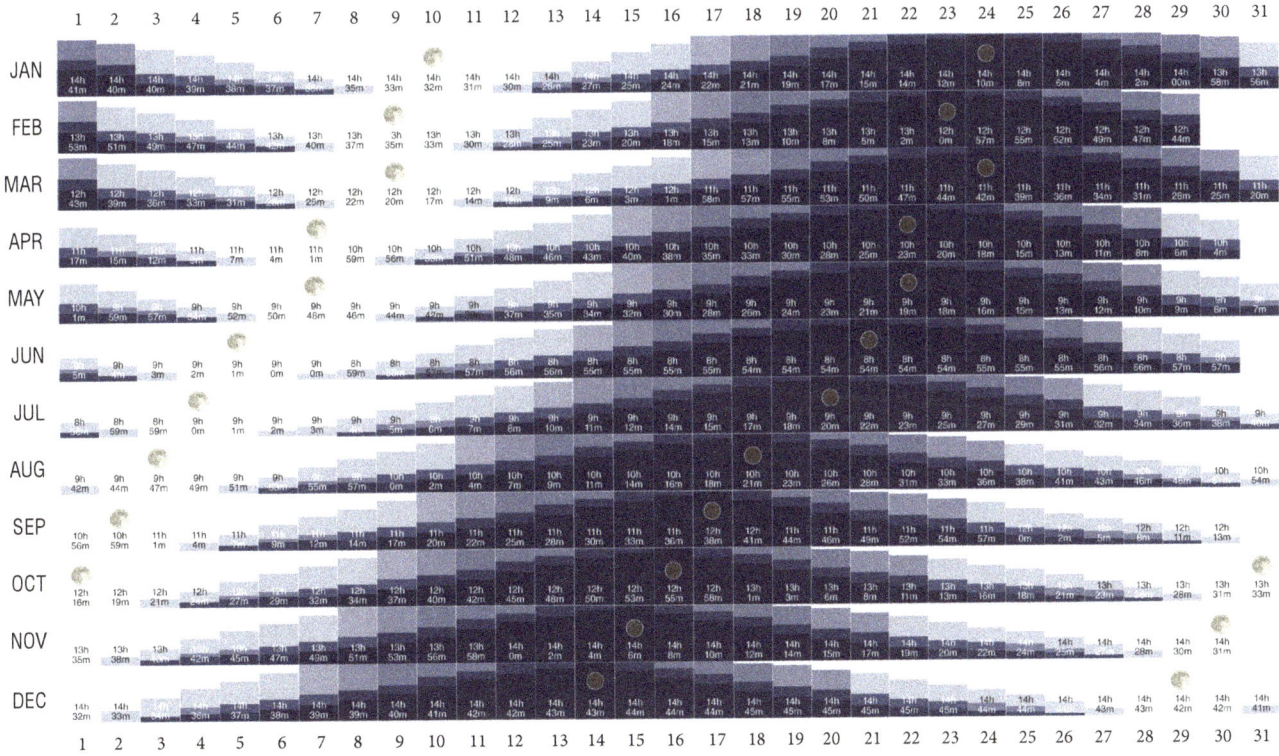

Legend:
- SUNRISE — Night ends on following day
- SUNSET — Nighttime begins
- Length of nighttime in hours (h) and minutes (m): 14h 36m / 14h 27m / 14h 36m
- No moon
- Relative illumination with moon above the horizon

	1	2	3	4	5	6	7	8	9	10	11	12	13	14	15	16	17	18	19	20	21	22	23	24	25	26	27	28	29	30	31
JAN	14h 41m	14h 40m	14h 40m	14h 39m	14h 38m	14h 37m	14h 36m	14h 35m	14h 33m	14h 32m	14h 31m	14h 30m	14h 28m	14h 27m	14h 25m	14h 24m	14h 22m	14h 21m	14h 19m	14h 17m	14h 15m	14h 14m	14h 12m	14h 10m	14h 8m	14h 6m	14h 4m	14h 2m	14h 00m	13h 58m	13h 56m
FEB	13h 53m	13h 51m	13h 49m	13h 47m	13h 44m	13h 42m	13h 40m	13h 37m	13h 35m	13h 33m	13h 30m	13h 28m	13h 25m	13h 23m	13h 20m	13h 18m	13h 15m	13h 13m	13h 10m	13h 8m	13h 5m	13h 2m	12h 0m	12h 57m	12h 55m	12h 52m	12h 49m	12h 47m	12h 44m		
MAR	12h 43m	12h 39m	12h 36m	12h 33m	12h 31m	12h 28m	12h 25m	12h 22m	12h 20m	12h 17m	12h 14m	12h 11m	12h 9m	12h 6m	12h 3m	12h 1m	11h 58m	11h 57m	11h 55m	11h 53m	11h 50m	11h 47m	11h 44m	11h 42m	11h 39m	11h 36m	11h 34m	11h 31m	11h 28m	11h 25m	11h 20m
APR	11h 17m	11h 15m	11h 12m	11h 9m	11h 7m	11h 4m	11h 1m	10h 59m	10h 56m	10h 53m	10h 51m	10h 48m	10h 46m	10h 43m	10h 40m	10h 38m	10h 35m	10h 33m	10h 30m	10h 28m	10h 25m	10h 23m	10h 20m	10h 18m	10h 15m	10h 13m	10h 11m	10h 8m	10h 6m	10h 4m	
MAY	10h 1m	9h 59m	9h 57m	9h 54m	9h 52m	9h 50m	9h 48m	9h 46m	9h 44m	9h 40m	9h 37m	9h 34m	9h 32m	9h 30m	9h 28m	9h 26m	9h 24m	9h 23m	9h 21m	9h 19m	9h 18m	9h 16m	9h 15m	9h 13m	9h 12m	9h 10m	9h 9m	9h 8m	9h 7m		
JUN	9h 5m	9h 3m	9h 2m	9h 1m	9h 0m	9h 0m	8h 59m	8h 57m	8h 56m	8h 56m	8h 55m	8h 55m	8h 55m	8h 54m	8h 54m	8h 54m	8h 54m	8h 54m	8h 54m	8h 55m	8h 55m	8h 55m	8h 56m	8h 56m	8h 57m	8h 57m					
JUL	8h 58m	8h 59m	8h 59m	9h 0m	9h 1m	9h 2m	9h 3m	9h 5m	9h 6m	9h 8m	9h 10m	9h 11m	9h 12m	9h 14m	9h 15m	9h 17m	9h 18m	9h 20m	9h 22m	9h 23m	9h 25m	9h 27m	9h 29m	9h 31m	9h 32m	9h 34m	9h 36m	9h 38m	9h 40m		9h
AUG	9h 42m	9h 45m	9h 47m	9h 49m	9h 51m	9h 53m	9h 55m	9h 57m	10h 0m	10h 2m	10h 4m	10h 7m	10h 9m	10h 11m	10h 14m	10h 16m	10h 18m	10h 21m	10h 23m	10h 26m	10h 28m	10h 31m	10h 33m	10h 36m	10h 38m	10h 41m	10h 43m	10h 46m	10h 48m	10h	10h 54m
SEP	10h 56m	10h 59m	11h 1m	11h 4m	11h 7m	11h 9m	11h 12m	11h 14m	11h 17m	11h 20m	11h 22m	11h 25m	11h 28m	11h 30m	11h 33m	11h 36m	12h 38m	12h 41m	11h 44m	11h 46m	11h 49m	11h 52m	11h 54m	11h 57m	12h 0m	12h 2m	12h 5m	12h 8m	12h 11m	12h 13m	
OCT	12h 16m	12h 19m	12h 21m	12h 24m	12h 27m	12h 29m	12h 32m	12h 34m	12h 37m	12h 40m	12h 42m	12h 45m	12h 48m	12h 50m	12h 53m	12h 55m	12h 58m	13h 1m	13h 3m	13h 6m	13h 8m	13h 11m	13h 13m	13h 16m	13h 18m	13h 21m	13h 23m	13h 28m	13h	13h 31m	13h 33m
NOV	13h 35m	13h 38m	13h 40m	13h 42m	13h 45m	13h 48m	13h 51m	13h 53m	13h 56m	13h 58m	14h 0m	14h 2m	14h 4m	14h 6m	14h 8m	14h 10m	14h 12m	14h 14m	14h 15m	14h 17m	14h 20m	14h 22m	14h 24m	14h 25m	14h	14h 28m	14h 30m	14h 31m			
DEC	14h 32m	14h 33m	14h 34m	14h 36m	14h 37m	14h 38m	14h 39m	14h 39m	14h 40m	14h 41m	14h 42m	14h 42m	14h 43m	14h 43m	14h 44m	14h 44m	14h 45m	14h 45m	14h 45m	14h 45m	14h 45m	14h 45m	14h 44m	14h 44m	14h 44m	14h 43m	14h 43m	14h 43m	14h 42m	14h 42m	14h 41m

2020 TIDES

The major influence creating tides is the gravitational effect of the Moon. If it were only a simple interaction, a high tide would occur as the Moon was at the meridian of a given location, its point of transit. But complex interactions with other factors generate a lag before a high or low tide peaks, called the lunitidal interval.

There is an average lunitidal interval for every location — if this number is known, the time of high tide can be predicted by subtracting the interval from the known time of the Moon's meridian.

Factors affecting the tidal cycle include the Sun's gravity, underwater and shoreline topography, the depth of water, the size of the body of water, and the weather. Barometric pressure alone is enough to raise or lower sea level by a foot or more.

The two phases in the lunar cycle when tides are highest are at the new moon and the full moon, when the gravitational forces of the Sun and Moon are aligned. These monthly high tides (and corresponding low tides) are called spring tides (from the German word *springen* — "to rise up"). The semi-high and -low tides that occur during the first and last quarter moons are called neap tides (from the Middle English word *neep* pertaining to such tides).

Extra-high tides called perigean spring tides occur during a full or new moon when the Moon is at its perigee, when it is closest to the Earth in its orbit (if this is the closest perigee of the year, the event is called a proxigean spring tide). The two events need to occur within a few hours of each other in order to enable a magnifying effect (the full or new moon has to occur within about five hours of the perigee to achieve the combined effect). Most of the time, monthly perigees do not coincide with full or new moons, making this a relatively rare event, occurring about every one and a half years. In the past 400 years, records show there were thirty-nine proxigean spring tides. Between 2000 and 2036, eight proxigean spring tides are predicted (*Journal of Coastal Research, March 2007*).

Spring tides are about 20 percent higher or lower than average tides and proxigean spring tides are even higher, but the excess is not always predictable. Historical records show that proxigean spring tides have sometimes, but not always, brought coastal flooding. Even with normal tides, weather conditions can completely mask tidal peaks — highs and lows — and delay or accelerate predicted tides by up to an hour, especially in coastal areas with relatively shallow water and sites that are more inland.

The next predicted proxigean spring tide is due on October 16, 2020. The new moon occurs at 2:31 PM EST and the perigee occurs at 7:00 PM EST, a difference of about five hours. Within the concept of a proxigean spring tide, this is a rather large gap, reducing the gravitational effects of the two events — Sun/Moon conjunction and closest point in the Moon's monthly orbit — working together. Close enough to pay attention, but not close enough to expect great tidal effects.

A long history of observations of tides and their effects suggests it is prudent to have conservative expectations from a proxigean spring tide. Offshore winds and storm surges usually create the most damage — regardless of any match-up with high tides — not the tide itself. An expected higher-than-normal tide could be a record-setting calamity or, like many such previous events of similar scale, turn out to be a dud. And as with most tide-related effects, the circumstances will vary from location to location.

The days of new moon and full moon — listed in the left and right columns — represent the days when the highest high and lowest low tides will occur. For a given location, the time of day or night for these tides will be the time of the Moon's local transit minus the lunitidal interval for this location. The most accurate times are to be found with location tide tables (in print or the online equivalent), or with special watches and clocks that can be set to produce these numbers.

Jan 2	11:45 PM EST	May 29	10:30 PM EST	Oct 23	12:23 PM EST
Feb 1	8:42 PM EST	Jun 28	3:16 AM EST	Nov 21	11:45 PM EST
Mar 2	2:57 PM EST	Jul 27	7:33 AM EST	Dec 21	6:41 PM EST
Apr 1	5:21 AM EST	Aug 25	12:58 PM EST		
Apr 30	3:38 PM EST	Sep 23	8:55 PM EST		

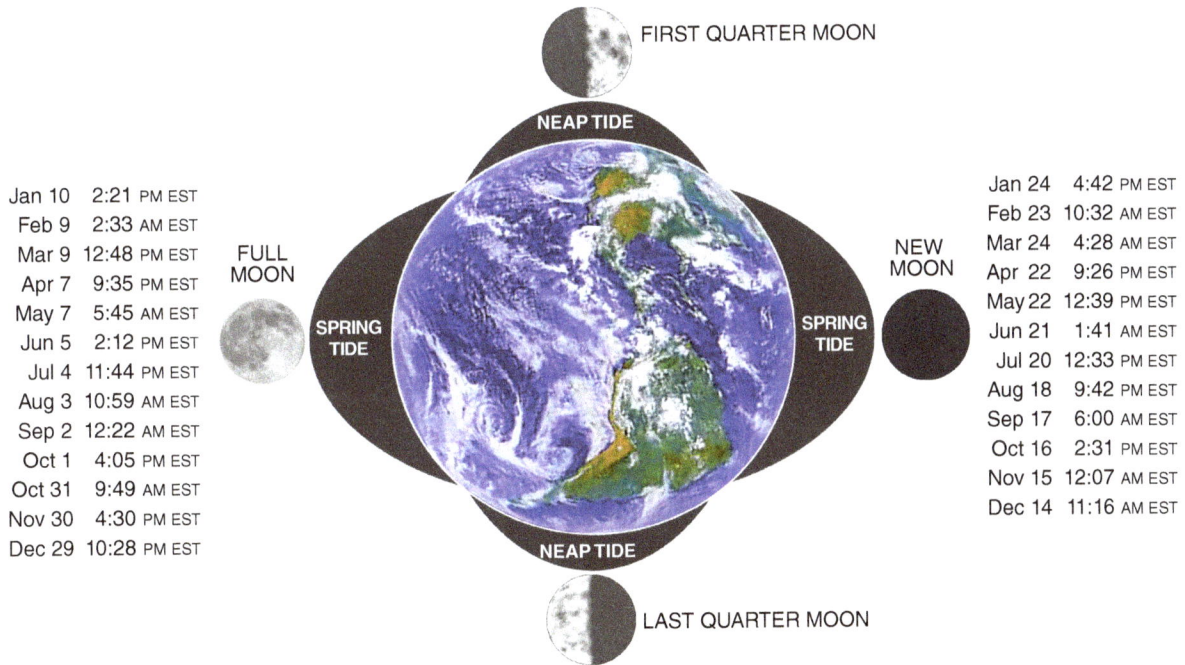

FIRST QUARTER MOON

NEAP TIDE

Jan 10	2:21 PM EST		Jan 24	4:42 PM EST
Feb 9	2:33 AM EST		Feb 23	10:32 AM EST
Mar 9	12:48 PM EST		Mar 24	4:28 AM EST
Apr 7	9:35 PM EST	FULL MOON / NEW MOON	Apr 22	9:26 PM EST
May 7	5:45 AM EST		May 22	12:39 PM EST
Jun 5	2:12 PM EST		Jun 21	1:41 AM EST
Jul 4	11:44 PM EST	SPRING TIDE / SPRING TIDE	Jul 20	12:33 PM EST
Aug 3	10:59 AM EST		Aug 18	9:42 PM EST
Sep 2	12:22 AM EST		Sep 17	6:00 AM EST
Oct 1	4:05 PM EST		Oct 16	2:31 PM EST
Oct 31	9:49 AM EST		Nov 15	12:07 AM EST
Nov 30	4:30 PM EST		Dec 14	11:16 AM EST
Dec 29	10:28 PM EST			

NEAP TIDE

LAST QUARTER MOON

Jan 16	7:58 PM EST	May 14	9:03 AM EST	Sep 10	4:26 AM EST
Feb 15	5:17 PM EST	Jun 13	1:24 AM EST	Oct 9	7:40 PM EST
Mar 16	2:34 AM EST	Jul 12	6:29 AM EST	Nov 8	8:46 AM EST
Apr 14	5:56 PM EST	Aug 11	11:45 AM EST	Dec 7	7:37 PM EST

MOON RESOURCES

SKY SIGHTS

This Week's Sky at a Glance (Sky & Telescope)
www.skyandtelescope.com

North American Skies
home.comcast.net/~sternmann/Welcome.html

EarthSky Tonight
www.earthsky.org

AstroPixels
www.astropixels.com

Guy's Blog
universalworkshop.com/guysblog

MOON ASTRONOMY

U.S. Naval Observatory
www.usno.navy.mil/astronomy

USGS Astrogeology Science Center
astrogeology.usgs.gov/SolarSystem/Earth/Moon

WHAT THE MOON LOOKS LIKE

aa.usno.navy.mil/idltemp/current_moon.php

MOON CALCULATORS & CYCLE DATA

aa.usno.navy.mil/faq/docs/moon_phases
aa.usno.navy.mil/data/docs/RS_OneYear.php

moon-phase-calculator.software.informer.com

TIDE PREDICTIONS

Tides & Currents (NOAA)
co-ops.nos.noaa.gov

ECLIPSES

aa.usno.navy.mil/data/docs/UpcomingEclipses.php
www.eclipsewise.com

CRESCENT MOON VISIBILITY

aa.usno.navy.mil/faq/docs/islamic.php
www.jgiesen.de/nmo/index.html

PLANETARIUM & SKY SOFTWARE

Stellarium
www.stellarium.org

Quick Phase Calculator
www.calculatorcat.com/moon_phases

Lunar Map Pro
www.riti.com/prodserv_lunarmappro.htm

SkyX/Software Bisque
www.bisque.com

APPS

Moon (iPhone, iPad)
itunes.apple.com

MoonPhase (iPhone, iPad)
itunes.apple.com

Deluxe Moon (Android, iPhone)
www.lifewaresolutions.com

Moon Watch (Android
www.paradex.co.uk

ORGANIZATIONS

American Association of Amateur Astronomers
www.astromax.com

American Astronomical Society
www.aas.org

Astronomical League
www.astroleague.org

National Space Science Data Center (NASA)
www.nssdc.gsfc.nasa.gov

Astronomical Society of the Pacific
www.astrosociety.org

British Astronomical Association
www.britastro.org

Royal Astronomical Society of Canada
www.rasc.ca

International Occultation and Timing Association
www.lunar-occultations.com

The International Planetarian Society
www.ips-planetarium.org

MAGAZINES

Amateur Astronomy Magazine
www.amateurastronomy.com

Astronomy Magazine
www.astronomy.com

Nature Magazine
www.nature.com

The Old Farmer's Almanac
www.almanac.com

Popular Astronomy Magazine
www.popastro.com

Sky & Telescope Magazine
www.skyandtelescope.com

Sky News
www.skynews.ca

CALENDARS & HOLIDAYS

Interfaith Calendar
www.interfaith-calendar.org

Time and Date
www.timeanddate.com

GLOSSARY

albedo. Percentage of light reflected from the Moon's surface: during a full moon, its albedo is 1.

angular diameter. The diameter of a distant object measured in degrees.

anomalistic month. The period of time it takes the Moon to go from one apogee (or perigee) to the next: 27.55455 days.

apogee. The point in the Moon's orbit when it is farthest from the Earth.

axis. An imaginary line through the center of mass of an object, around which the object rotates.

azimuth. The angle measured from an observer's due north to directly under an object of interest in the sky. With North at 0°, the numerical points move clockwise: East is 90°, South is 180°, and West is 270° — 360° is the same as 0°.

conjunction. The position of two celestial bodies when they are in line with one another as seen by an observer on Earth. The new moon can be referred to as the Moon in conjunction with the Sun (opposite of opposition).

crescent moon. The Moon's phase just before and after the new moon, when only a thin curved section is lighted by the Sun.

culmination. The highest point a celestial body reaches in the sky as seen from Earth, always occurring when the body's azimuth is 180, or due south.

dark of the moon. Another name for new moon.

declination. The angle measured between the celestial equator and an object in the sky.

earthshine. Light from the Earth reflected on the Moon, visible during lunar eclipses and young crescent moons. This effect is often referred to as the "old moon in the new moon's arms."

eclipse. The blocking of light occurring when the Moon or Earth blocks sunlight. *Lunar eclipse*: the shadow of the Earth blocks the Moon; can only occur during a full moon. *Solar eclipse*: the Moon blocks the Sun; can only occur during a new moon. *Annular eclipse*: the Moon is farthest away in its orbit, leaving a ring of light around the Moon's shadow. *Partial eclipse* (penumbral eclipse): the Earth's shadow does not completely cover the Moon.

ecliptic. The imaginary plane formed by the Earth's orbit around the Sun; the apparent path of the Sun.

elongation. The angle of a planet away from the Sun or the Moon as viewed from the Earth.

first quarter moon. Moon's phase when it is 90 degrees away from a line between the Sun and Earth. The angle of illumination creates a half-circle image on the Moon's surface, with the lighted half on the right.

full moon. Moon's phase when it is on the opposite side of the Earth from the Sun and receives sunlight across its entire face, forming a full circle of light.

gibbous moon. Period when the Moon is getting larger after the first quarter moon (waxing gibbous) or smaller after the full moon but before the last quarter moon (waning gibbous).

last quarter moon. Moon's phase when it is 90 degrees away from a line between the Sun and the Earth. The angle of illumination creates a half circle on the Moon's surface, with the lighted half being on the left.

librations. Irregular motions of the Moon in its elliptical orbit around Earth that allow slightly more than half of the surface to be visible over a period of time.

lunar day. The period of time between two successive transits of the Moon over the same meridian: 24.84 hours.

meridian. An imaginary line that passes directly north and south through an observer or specified location on Earth. A plane extended from this line into space passes through the zenith (point above the observer).

moonrise/moonset. The point in time when the upper limb of the Moon is even with the Earth's horizon as the Moon rises in the east or sets in the west.

Metonic Cycle. Every 19 years (approximately), the lunar phases line up with the calendar months, a pattern first noted in Europe in 432 B.C.

neap tide. The lowest high tide in a lunar month, occurring near the first and last quarter moon phases.

new moon. Moon's phase when it is directly between the Earth and the Sun. Because sunlight is hitting only the far side of the Moon, it appears dark from the Earth.

occultation. The movement of one celestial object behind another.

opposition. A specific point in time when a moon or planet is 180 degrees away from the Sun relative to the Earth. The Moon is full when it is in opposition to the Sun.

penumbra. The light part of a shadow formed by diffused light in an area around the edges of an object.

perigee. The point in the Moon's orbit when it is closest to the Earth (opposite of apogee).

perigean tide. The high tide of the month that occurs when the Moon is closest to Earth.

quadrature. The position of the Moon or a planet when it is at right angles to the Sun. The Moon is in first quarter phase when it is in east quadrature to the Sun and last quarter phase when it is in west quadrature.

Saros Cycle. A cycle of lunar months lasting 18 years, 11.3 days, the time it takes the Moon, the Earth, and the Sun to return to the same relative position.

sidereal month. A lunar month measured by a return to the position marked by a specific star: 27.32166 days.

spring tide. The highest tide in a lunar month, occurring near new and full moons.

synodic month. A lunar month as measured from the point of one new moon to the next new moon: 29.53059 days.

terminator. The line formed at the edge of the illuminated portion of the Moon.

transit. The point when the path of the Moon, the Sun, a star, or a planet takes it across the meridian.

umbra. The darker core of a shadow, usually cone shaped, (surrounded by a lighter penumbral shadow). Also refers to the darker center of sunspots.

universal time. The time zone centered at 0° longitude, the position of the Greenwich Observatory in England (traditionally known as Greenwich Mean Time).

waning moon. The period after the full moon and before the new moon; the lighted portion of the Moon's surface is decreasing.

waxing moon. The period after the new moon and before the full moon; the lighted portion of the Moon's surface is increasing.

zenith. The imaginary point directly above an observer on Earth.

www.ingramcontent.com/pod-product-compliance
Lightning Source LLC
Chambersburg PA
CBHW051802200326
41597CB00025B/4654